T0191832

Sustainable Textiles: Production, Processing, Manufacturing & Chemistry

Series Editor

Subramanian Senthilkannan Muthu, Head of Sustainability, SgT and API, Kowloon, Hong Kong

This series aims to address all issues related to sustainability through the lifecycles of textiles from manufacturing to consumer behavior through sustainable disposal. Potential topics include but are not limited to: Environmental Footprints of Textile manufacturing; Environmental Life Cycle Assessment of Textile production; Environmental impact models of Textiles and Clothing Supply Chain; Clothing Supply Chain Sustainability; Carbon, energy and water footprints of textile products and in the clothing manufacturing chain; Functional life and reusability of textile products; Biodegradable textile products and the assessment of biodegradability; Waste management in textile industry; Pollution abatement in textile sector; Recycled textile materials and the evaluation of recycling; Consumer behavior in Sustainable Textiles; Eco-design in Clothing & Apparels; Sustainable polymers & fibers in Textiles; Sustainable waste water treatments in Textile manufacturing; Sustainable Textile Chemicals in Textile manufacturing. Innovative fibres, processes, methods and technologies for Sustainable textiles; Development of sustainable, eco-friendly textile products and processes; Environmental standards for textile industry; Modelling of environmental impacts of textile products; Green Chemistry, clean technology and their applications to textiles and clothing sector; Eco-production of Apparels, Energy and Water Efficient textiles. Sustainable Smart textiles & polymers, Sustainable Nano fibers and Textiles; Sustainable Innovations in Textile Chemistry & Manufacturing; Circular Economy, Advances in Sustainable Textiles Manufacturing; Sustainable Luxury & Craftsmanship; Zero Waste Textiles.

More information about this series at https://link.springer.com/bookseries/16490

Subramanian Senthilkannan Muthu
Editor

Sustainable Approaches in Textiles and Fashion

Circular Economy and Microplastic Pollution

Editor
Subramanian Senthilkannan Muthu
Head of Sustainability
SgT and API
Kowloon, Hong Kong

ISSN 2662-7108 ISSN 2662-7116 (electronic)
Sustainable Textiles: Production, Processing, Manufacturing & Chemistry
ISBN 978-981-19-0532-2 ISBN 978-981-19-0530-8 (eBook)
https://doi.org/10.1007/978-981-19-0530-8

This Springer imprint is published by the registered company Springer Nature Singapore Pte Ltd.
The registered company address is: 152 Beach Road, #21-01/04 Gateway East, Singapore 189721, Singapore

Contents

Editor and Contributors

About the Editor

Dr. Subramanian Senthilkannan Muthu currently works for SgT Group as Head of Sustainability and is based out of Hong Kong. He earned his Ph.D. from the Hong Kong Polytechnic University and is a renowned expert in the areas of Environmental Sustainability in Textiles & Clothing Supply Chain, Product Life Cycle Assessment (LCA) and Product Carbon Footprint Assessment (PCF) in various industrial sectors. He has five years of industrial experience in textile manufacturing, research and development and textile testing, and over a decade of experience in Life Cycle Assessment (LCA), carbon and ecological footprints assessment of various consumer products. He has published more than 100 research publications, written numerous book chapters and authored/edited over 100 books in the areas of carbon footprint, recycling, environmental assessment and environmental sustainability.

Contributors

S. Raja Balasaraswathi Department of Fashion Technology, PSG College of Technology, Coimbatore, India

D. G. K. Dissanayake Department of Textile and Apparel Engineering, University of Moratuwa, Katubedda, Sri Lanka

Minakshi Jain Government Girls College, Chomu, Rajasthan, India

G. Jeya Research Department of Chemistry, Pachaiyappa's College, Chennai, India

M. S. Parmar Northern India Textile Research Association, Rajnagar, Ghaziabad, India

Shanthi Radhakrishnan Costume and Apparel Designing, PSGR Krishnammal College for Women, Coimbatore, India

R. Rathinamoorthy Department of Fashion Technology, PSG College of Technology, Coimbatore, India

Nidhi Sisodia Northern India Textile Research Association, Rajnagar, Ghaziabad, India

V. Sivamurugan Research Department of Chemistry, Pachaiyappa's College, Chennai, India

V. Sivasankar Research Department of Chemistry, Pachaiyappa's College, Chennai, India

B. Subathra Department of Apparel and Fashion Design, PSG College of Technology, Coimbatore, India

T. G. Sunitha Research Department of Chemistry, Pachaiyappa's College, Chennai, India

J. Swetha Jayalakshmi Department of Apparel and Fashion Design, PSG College of Technology, Coimbatore, India

D. Vijayalakshmi Department of Apparel and Fashion Design, PSG College of Technology, Coimbatore, India

Synthetic Textile and Microplastic Pollution: An Analysis on Environmental and Health Impact

S. Raja Balasaraswathi and R. Rathinamoorthy

Abstract The term 'fast fashion' has become the most fascinating term that is getting greater acceleration in the global fashion market. It aims to make the quick arrival of high-end fashion to the mass market at affordable prices. The business of the fashion industry is getting promoted to a huge level due to the fast-fashion system; however, the system fails to address sustainability. Though the business and profits are focused on the one end, the other side seeks sustainability. The fashion and apparel industries have started to adopt eco-friendly trends to move toward sustainability in the industry. While the aim is to achieve a sustainable wardrobe, it is not only limited to bringing up sustainable materials to the wardrobe but also to address the issue with the things which are already there. The synthetic fabrics which are dominant in the fast fashion trend (around 60%) have the problem of shedding microfibers that can add on to the microplastic pollution. The domestic washing of synthetic garments leads to the generation of microfibers into the marine environment. It has been estimated that around 22 million tons of microfibers will be entering the marine environment by the year 2050. These microfibers in the marine environment can be mistaken as food by various living organisms and the issue got more concern when it has the possibility to end up in human beings through the food chain. Moreover, microfibers are being found in seafood, table salt, drinking water, and even in the air which can cause an adverse impact on human beings. This chapter addresses the arising impacts of microfiber pollution in the environment and the importance of seeking a solution for the issue. It also aims to elaborate the need for research from the textile and fashion aspirants addressing the issue to make a way for the industry to move toward sustainability.

Keywords Fast Fashion · Sustainability · Microplastics · Synthetic textiles · Microfiber shedding · Environmental Impact · Food Chain · Health Hazards

---

S. R. Balasaraswathi · R. Rathinamoorthy (✉)
Department of Fashion Technology, PSG College of Technology, Coimbatore, India

S. S. Muthu (ed.), *Sustainable Approaches in Textiles and Fashion*, Sustainable Textiles: Production, Processing, Manufacturing & Chemistry,
https://doi.org/10.1007/978-981-19-0530-8_1

1 Introduction

Sustainability has become the most coveted thing in every field all over the world in which textile and fashion industry is not an exception. Textile and fashion industries are consuming a huge number of natural resources in the name of raw materials [1]. Nearly 70 million barrels of oil per year are used for polyester fiber production [2], whereas various processes involved in the industry require huge energy and manpower and can emit hazardous waste into the environment [1]. The fashion industry is responsible for 20% of global wastewater and 10% of the world's carbon emissions [3]. Moreover, the demand of the industry increases with the fast-fashion system which is prevailing in the market in recent years. The system accelerates the global fashion market and results in an increased volume of business. The huge production of apparel in a shorter period at an affordable price to make the latest trends available for a large volume of people is the strategy of the fast-fashion system [1]. Though the system increases business and profits [4], the sustainability aspect of the system has become a challenging one. The increased production results in huge depletion of resources, whereas the rapid changes in the trend result in underutilization of garments and increased disposal rates. Garment consumption has raised by 400% in the past twenty years [5]. With this much increase in demand, it has been estimated that by the year 2050, the industry would be in the need of 3 times more resources than in 2000 [6]. The demand is being overestimated to avoid losing a large volume of business which results in unsold stocks. In 2018, H & M has announced that stocks worth 4.3 billion dollars remained unsold [7]. Moreover, the shorter trend cycle and lower price make the garments be disposed of effortlessly. It has been estimated that an average American can produce 82 pounds of textile waste every year [5]. Around 85% of textiles produced got dumped every year [6]. This textile waste can adversely affect the environment. Figure 1 highlights the strategies of fast fashion with their benefits and drawbacks.

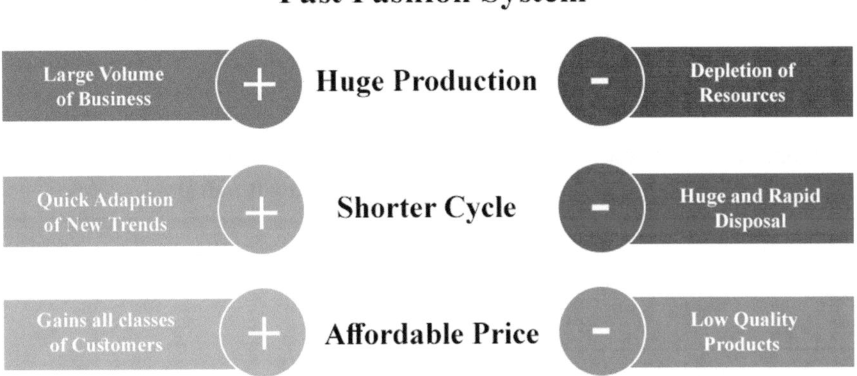

Fig. 1 Benefits and drawbacks of fast-fashion system

Synthetic fibers, especially polyester, are the most dominant fibers (around 60%) in the fast-fashion system because of their unbeatable advantages over other materials [8]. Synthetic fibers are friendly to manufacturers because of their low cost [9] and easy production [10], whereas they seek end-users' attention by extraordinary properties [9]. On the other hand, the problems associated with synthetic materials are unmanageable. The production of synthetic textiles depends on non-renewable resources [2]. In the production stage, the fiber production reaction results in certain toxic by-products like nitrous oxide [2], antimony, and ethylene glycol (carcinogens) [3]. Moreover, in the dyeing process, synthetic textiles require hazardous chemicals for proper dye fixation. Around 5–15% of disperse dyes that are used for synthetic textile dyeing end up in a freshwater system [3]. The production and usage of synthetic textiles greatly contribute to carbon emissions. The production of synthetic clothing also results in a huge volume of carbon dioxide emission which is equivalent to 20 tons of CO_2 per person in the world [2]. To make the situation worse, the wearing of synthetic textiles could cause an adverse effect. The residual monomeric forms on the fibers can penetrate the human body through the skin. Prolonged wearing of polyester can cause chronic respiratory infections [11].

While many research works have been made on different sustainability issues associated with synthetic textiles, some of the risks associated with synthetic textiles were not addressed effectively. One of such risks with synthetic textiles is microplastic pollution. The plastic market has an unbeatable place in the global market with a value of USD 579.7 billion in the year 2020. With a 3.4% compounded annual growth rate, the market size is expected to expand greatly [12]. The plastic particles whose size is not more than 5 mm are considered microplastics. With the growing demand for plastics, the prevalence of microplastics is also getting increased. Plastics are being used in various industries, and in textile industries, plastics are used in the form of synthetic textiles. In the year 2016, around 20% of total plastics produced were synthetic textile fibers which are roughly equivalent to 65 million tons [13]. The demand for polyester fibers has increased exponentially than other natural fibers in the past 20 years, and it is expected that synthetic textiles will share 75% of total apparel fiber production by the year 2030 [14]. It has been believed that the fashion industry accounts for 31% of total plastic pollution in the ocean [6]. Around 8 million tons of plastic wastes are entering the marine environment, out of which 1.5 million tons are microplastics [15]. Polyester and other synthetic textiles can potentially account for microplastics in the form of microfibers [16]. The increased demand and consumption of synthetic textiles which are accelerated due to the fast-fashion system increase the contribution of synthetic textiles in microplastic pollution. It has been estimated that between 2015 and 2050, two-third of synthetic fibers which are being used for apparel production presently will be entering into the ocean [13]. With the increasing magnitude of microplastics in different environments, recently environmentalists have done researches on the prevalence of microplastics, their possible sources, and potential risks associated with them. This chapter aims to highlight the contribution of synthetic textiles to microplastic pollution and their environmental as well as human health impacts. The later part discusses the research status of microfiber pollution which provides a clear insight on the research gap in this area which can potentially support future research to control microfiber pollution.

2 Ubiquity of Microplastics

Microplastic pollution has become a threat to the world after the prevalence of these micro-sized particles in different levels of the environment has been witnessed. Microplastic particles have been detected in the aquatic environment, terrestrial environment, and even in the atmosphere. This section provides an insight into the current status of microplastic pollution in the different environments reported earlier.

2.1 Microplastics in Aquatic Environment

Microplastics were found in the marine environment as well as freshwater systems. Various aquatic regions including surface water, deep-sea sediment, and coastal lines have been examined by researchers for microplastic contaminants. Zhao et al. investigated the surface water of Yangtze Estuary in China. They have reported that their sampling sites were contaminated with microplastic particles, and they have quantified the average number of microplastics as 4137.3 $\pm$ 2461.5 particles per meter cube of surface water [17]. In line with this, the surface water of the Eastern Indian Ocean is found to be contaminated with microplastic particles. Studies have claimed a higher level of polypropylene followed by polyethylene, phenoxy resin, polystyrene, polyester, polyamide, polyacrylonitrile, polyvinylchloride, polyether urethane, and polyethylene terephthalate [18]. While analyzing the deep-sea sediments of Southern Europe, the research findings showed both natural and synthetic microfibers. They observed a huge level of regenerated cellulose followed by polyester and acrylic fibers. While studying in depth the source of contamination, a direct link between the type of clothes washed in domestic laundry and the fiber contamination in the deep sea has been noted [19]. The other study has noticed microfibers in Antarctica Bay and has also added that the contamination in such places where human presence is lesser is due to the increased plastic waste in oceans that are brought by the Bransfield current [20].

Similarly, the freshwater systems including rivers and lakes were reported with microplastics. Researchers who explored the sediments of St. Lawrence River (North America) have found microbeads of different sizes and colors in the sediments. They have claimed these microbeads as polyethylene-composed particles with their melting point at 113.7 °C [21]. In line with this, other studies have revealed microplastic contamination of rivers in the Tibet Plateau. They have examined surface water as well as sediments and reported 483–967 microplastic particles per cubic meter of surface water, whereas 50–195 microplastic particles per kilogram of sediment. Moreover, they have confirmed the chemical composition of the particles using micro-Raman spectroscopy, which shows polyethylene terephthalate, polyethylene, polypropylene, polystyrene, and polyamide as commonly found polymers [22]. Similarly, the river bed (different depths) of the Rhine River was analyzed by Mani et al., and it has been proclaimed that the microplastic contamination decreases with an increase in depth. They have examined particles in the size range of 11–500 μm and identified 18 different types of polymers that include chlorinated polyethylene, ethylene-propylene-diene rubber, polyester, and polyethylene [23].

The Laurentian Great Lakes was investigated for plastic debris. With a 333-μm mesh trawl, they analyzed the surface water of the lake and reported 43,000 microplastic particles per square kilometer [24]. In line with this, Egessa et al. recorded microplastics in Lake Victoria, the largest lake in Africa. They have mentioned that the microplastic particles were the disintegrated forms of larger plastics. They have added that the particles identified were of low-density polymers including polyethylene and polypropylene [25]. The rainwater pipelines and rainwater sediments studies revealed that they have huge potential to transport the microplastics to the freshwater systems. The number of microplastic particles per liter ranges between 2.75 ± 0.76 and 19.04 ± 2.96, in which the most commonly noted morphology of particles is plastic fragments and fibers [26].

From these research reports, it is evident that the prevalence of microplastics in the aquatic system is not negligible. The major sources of contamination of the aquatic environment as identified by the researchers include washing of synthetic textiles [19] in the case of water bodies nearer to human habitats, whereas leaching of degraded plastics from different sources [20, 25, 27] can be responsible for contamination of sites where human residency is lesser.

2.2 *Microplastics in Terrestrial Environment*

The prevalence of microplastic particles in the terrestrial environment has been reported by various researchers. Ambrosini et al. have identified the microplastic contamination of terrestrial glaciers. They have analyzed one of the widest Italian glaciers with a surface area of 11.34 km^2 and revealed around 74.4 ± 28.3 items per kilogram of sediment. They have also estimated 570–801 million microplastic particles in one square kilometer of the ablation region of the glacier. The clothes and other equipment used by the tourists visiting the glacier could be the possible answers for the contribution of these microplastic particles. The prevalence of these microplastic particles in the glaciers also has the potential to enter the freshwater as well as marine environment [27].

When the vegetable fields around Shanghai were inquired, shallow and deep soil contaminations were reported. They have mentioned shallow soil (78.00 ± 12.91 microplastic particles/kg) to be more contaminated than that of deep soil

(62.50 ± 12.97 microplastic particles/kg) [28]. Similarly, Chen et al. recorded 320–12,560 microplastic particles per kilogram of vegetable farmlands in Wuhan. The suburban roads and residential areas around the farmland were considered as the source of microplastic particles which can be brought to farmlands through wastewater discharge [29]. This might increase the chances of contamination of vegetables. The other researchers have mentioned plastic mulching as the potential source of microplastic in the terrestrial environment [30]. Mulching is the process of covering land/soil to provide favorable environment for plant growth. Plastic mulch can accelerate healthy plant growth by providing increased soil temperature, reduced evaporation, and increased nutrients [31]. This is supported by the microplastic abundance in the soil where plastic mulching done for 15 years is significantly greater than the soil where mulching is done for 5 years [30].

Sewage sludge has been noted as one of the major sources of terrestrial contamination. The sewage treatment plants filter the microplastic wastes and are left in sludge which ends up in the terrestrial environment. Sewage sludges are generally used as fertilizer for the soil. Corradini et al. evaluated the sludge samples and the soils where sludges were applied. They have reported an average of 34 particles per gram of sludge [32]. On the sewage sludge collected from different sewage treatments across the United States, polyethylene terephthalate (370 μg/g) and polycarbonate (5.9 μg/g) were identified in the sludge. Based on their analysis, the researchers estimated that around 4,010,000 kg of microplastics will be entering the terrestrial environment every year through sewage sludges [33].

The terrestrial environment is also found to be contaminated as well as aquatic systems. A wide range of microplastic polymers was found in terrestrial glaciers and soils [27, 29, 30]. The deposition of sewage sludge and discharge of sewage effluents into fields were found as the dominant source of microplastics in farmlands [29, 32], whereas plastic mulching also contributes to an extent [30].

2.3 *Microplastics in Atmosphere*

Studies have confirmed the prevalence of microfibers in the atmosphere. The researchers who have analyzed the prevalence of microplastics in the atmosphere over the East Indian Ocean and South China sea have reported an average of 1.0 (with a range of 0–7.7 particles) microplastic particles per 100 m^3 of the atmosphere. The nearby terrestrial microplastics are contributing to the microplastic particles in the atmosphere, whereas the transport of microplastic particles through the atmosphere is the key reason for the microplastics in the atmosphere which is far away from the land-based sources. They have also mentioned that the distribution of microplastic particles in the atmosphere can be varied with pressure, wind speed, humidity, and gust velocity [34]. In similar research, microplastic particles have been found in the atmospheric air of Shanghai. An average of 1.42 ± 1.42 particles per cubic meter of air was noted. They have identified polyethylene terephthalate as the dominant fiber. They have added that the prevalence of microplastics in the atmosphere urges the analysis of the exposure of human beings to microplastics through inhalation besides ingestion [35]. The microplastic contamination of indoors and outdoors has been measured and the average number of microplastic particles is higher in the indoor atmosphere (3–15 particles/cu.m) than that of outdoor atmosphere (0.2–0.8 particles/cu.m). This is due to the high deposition velocity of microplastic particles in the outdoor atmosphere [36]. The other researchers evaluated the air in crowded areas (university campus and intercity) for microplastic contamination. They have observed microplastic fibers as well as fragments [37]. Table 1 summarizes the level of microplastic contamination of different environments.

3 Impact of Microplastic Pollution

3.1 Microplastics in the Food Chain

Microplastic particles were found in different edible items and in the various living organisms which are resulted from the consumption of contaminated items as well as direct consumption of microplastic particles by mistaking them as food. Studies have confirmed that microplastic particles in various aquatic organisms can end up in human beings through the food chain. Researchers have investigated blue mussels and lugworms for microplastic contamination. They have reported 0.1 ± 0.2 particles and 0.3 ± 0.6 particles per gram of mussel and lugworm tissues, respectively. Their analysis of the adverse effect of microplastics in these organisms has resulted in no significant difference in the metabolism of the organism after exposure to microplastics [38]. Microplastics were detected in the intestine of fishes in the Northern Bay of Bengal. Three different species, namely pink Bombay-duck (*Harpadonnehereus*), white Bombay-duck (*H. translucens*), and gold-stripe sardine (*Sardinella gibbosa*), were taken into study. The plastic particles in the gastrointestinal tract (0.37–1.55 items/g) of the fishes are dominant than the particles in the fish body (0.07–0.08 items/g) [39]. In a study performed on Australian seafood, the edible portions of oysters, prawns, squid, crabs, and sardines were evaluated. The pyrolysis gas chromatography and mass spectrometry method has confirmed microplastics in all the species examined with sardines as the most contaminated [40]. Microplastics were also noticed in different land-based animal species. Research on a wide variety of medicinal animals (20 different species) has revealed their contamination with microplastic particles. It clearly shows that the presence of microplastic particles is attributed to their ingestion through the food chain as the microplastic particles were found in the intestine. Moreover, this shows that besides aquatic organisms, land-based organisms were also contaminated with microplastic particles which increases the seriousness of the transfer of microplastic particles to human beings [41].

Though the microplastic particles are tracked down in the food chain and their ingestion by living beings, their impact is not clearly known. Researchers have identified the effect of microfibers (polyester fibers) on the soil invertebrates and have

Table 1 Prevalence of microplastic particles

Reference	Contaminated area	Place/location of research	Quantified microplastic particles
Aquatic environment			
Absher et al. [20]	Bay	Antarctica	2.40 ± 4.57microfibers/cu.m
Zhao et al. [17]	Estuary	China	4137.3 ± 2461.5 items/cu.m
Sanchez-Vidal et al. [19]	Deep sea	Southern Europe	10 – 70 microfibers/50 ml of sediment
Li et al. [18]	Surface water	Eastern Indian Ocean	0.34 ± 0.80 item/sq.m of Indian Ocean
Castañeda et al. [21]	River	North America	13,832 ± 13,677 items/sq.m
Mani et al. [23]	River	Europe	0.26 ± 0.01–$11.07 \pm 0.6 \times 10^3$ items/kg of sediment
Eriksen et al. [24]	Lake	North America	43,000 items/sq.km
Sang et al. [26]	Rain water pipelines and sediments	China	2.75 ± 0.76 – 19.04 ± 2.96 items/L of water 6.00 ± 1.63 – 27.33 ± 4.64 items/100 g of sediment
Egessa et al. [25]	Lake	Africa	0.02–2.19 items/cu.m
Jiang et al. [22]	River	Tibet Plateau	483–967 items/cu.m of surface water 50–195 items/kg of sediment
Terrestrial environment			
Ambrosini et al. [27]	Glaciers	Italy	74.4 ± 28.3 items/kg of sediment
Chen et al. [29]	Vegetable farmland	China	320 to 12,560 items/kg of soil
Huang et al. [30]	Agricultural soil	China	80.3 ± 49.3 – 1075.6 ± 346.8 items/kg of soil
Corradini et al. [32]	Soil with sewage sludge	Chile	0.6–10.4 items/g of soil

(continued)

Table 1 (continued)

Reference	Contaminated area	Place/location of research	Quantified microplastic particles
Zhang et al. [33]	Sewage sludge	United States	28–12,000 μg of PET items/g 0.70–8400 μg of PC items/g
Liu M., et al. [28]	Farmlands	China	78.00 ± 12.91 items/kg of shallow soil 62.50 ± 12.97 items/kg of deep soil
Atmosphere			
Wang et al. [34]	Ocean-Atmosphere	Indian Ocean, South China Sea	1.0 items/cu.m of air
Gasperi et al. [36]	Indoor and outdoor atmosphere	France	3–15 particles/cu.m in indoor 0.2–0.8 particles/cu.m in outdoor
Liu, K., et al. [35]	Outdoor/city atmosphere	China	1.42 ± 1.42 items/cu.m of air

noticed very mild or no effect due to the ingestion of short fibers. However, they have observed a decrease in reproduction due to the long fibers as they are difficult to be ingested. They have added that the impact of these microfibers is not limited to the ingestion of these particles as the presence of these particles can change the surrounding of invertebrates and they can potentially harm physically [42]. The other researcher examined the effect of microplastic fibers on freshwater daphnids. They have noticed an increase in mortality after exposure to microplastics for 48 h [43].

The microplastics were also retrieved in different food items which include table salt, beer, and honey. A study that explored the commercial table salts in Taiwan has reported an average of 9.77 microplastic particles per kilogram of salt which includes different types of polymers such as polypropylene, polyethylene, polystyrene, polyester, polyetherimide, polyethylene terephthalate, and polyoxymethylene [44]. In line with this, the other research work that analyzed 16 table salt brands over 8 countries has noted microplastic particles with polypropylene, polyethylene, and polyethylene terephthalate as the most common polymer types [45]. In a study on milk, beer, honey, and soft drinks for microplastic contamination, microplastic fragments as well as fibers were detected. From their results, they have claimed a higher amount of microplastics in craft honey (67 items/liter) which is followed by industrial honey (54 items/liter), industrial beer (47 items/liter), skim milk (40 items/liter), and soft drinks (32 items/liter) [46]. Out of 23 different branded milk samples utilized in a study including a kids' milk brand, microplastic contamination was reported with fibrous material of blue color as the dominant one [47]. Similarly, microplastics in fruits and vegetables were also identified [48], which might be the result of microplastic contaminated farmlands [28]. Apple, pear fruit, potato, carrot, lettuce, and broccoli were examined for microplastic contamination, and the results showed apple and carrot as the most contaminated fruit and vegetable, respectively [48]. This much prevalence of microplastics in food items increases the threatening of microplastic exposure to human beings.

3.2 *Microplastics and Human Health*

Recently, studies have also explored the intake of microplastic particles by human beings. The presence of microplastics in various food items and the atmosphere has confirmed the exposure of human beings to microplastic particles through ingestion and inhalation. Research has estimated the average amount of microplastic particles intaken by Americans based on their diet. They have revealed that 39,000–52,000 particles are being consumed per year which varies based on the age and sex of the human. While considering the microplastics in the atmosphere (inhalation), this estimation can rise to 1,21,000 particles. They have highlighted that male adults are consuming more microplastics (312 particles/day) followed by female adults (258 particles/day), male children (223 particles/day), and female children (203 particles/day). They have noticed a huge difference in the microplastic consumption between the people consuming bottled drinking water (90,000 particles/year in addition to food intake and inhalation) and tap water (4,000 particles/year in addition to food intake and inhalation). And hence, they have suggested avoiding bottled drinking water to reduce exposure to microplastic particles [49]. Meanwhile, the study that examined microplastic particles in table salt has estimated the average consumption of microplastics by human beings through table salt and has estimated that around 525.75 microplastic particles/year are being consumed by humans [44]. Researchers who confirmed the presence of suspended atmospheric microplastics in Shanghai have proposed a model to estimate the exposure of human beings to microplastic particles. With that model, it has been estimated that around 21 microplastic particles could be possibly inhaled by humans in the outer environment of Shanghai [35].

Moreover, the microplastic particles were also observed in human tissues which confirms human exposure to microplastics. Researchers have observed microplastic particles in the maternal side, fetal side, and chorioamniotic membranes of the human placenta. They have highlighted that these particles may trigger immune responses which need to be detailed [50]. The other analysis sampled human hair, face, skin, and saliva for microplastic contamination. A higher abundance of microplastics in human hair has been noted, and males were found to be exposed to a higher level of microplastics than females. The sources of microplastics were home furnishing and wearing of clothes in the indoor scenario, whereas the deposition of atmospheric microplastics was noticed in the outdoor scenario [51]. Though the exposure of human beings to microplastics is confirmed, the effect of these microplastic particles on human beings is not analyzed. However, a study has been made to analyze the effect of polystyrene microplastics on human lung cells. The researchers have observed a significant effect on cell morphology. Moreover, the microplastic particles inhibit cell proliferation [52].

4 Microfibers—Major Source of Microplastics

Among different sources of microplastics including tires, city dust, road markings, marine coatings, personal care products, and synthetic textile materials have been identified as the most common and dominating source (35%) of microplastics [53]. Synthetic textiles can contribute to microplastics in the form of microfibers. The phenomenon of the detachment of loose or broken or damaged fibers from the surface of textile materials can be termed as microfiber shedding. The detachment may occur due to the external force application or inherent nature of textile materials, and release into the environment [54]. This can happen throughout the life cycle (production, consumption, and disposal) of the textile material [55].

As a support to the huge contribution of synthetic textile materials in microplastic pollution, various works characterized the microplastic particles based on their size and appearance and fibrous materials were identified as a superior contaminant. Researchers who examined microplastics in the glaciers have reported that 65.2%

of particles were of fibrous materials. Of those, the most dominating polymer is polyester followed by polyamide, polyethylene (PET), and polypropylene (PP) [27]. The atmosphere over oceans was also consisting of microplastic particles, out of which 88.89% were fibrous and 11.11% were fragments. PET used in the textile industry was the most dominant, which is followed by PP being used in textile, packing, and reusable materials. They have added that the dropout and breakdown of synthetic clothes could be the reason for the higher level of PET fibers in the atmosphere [34]. In a similar research where the atmospheric microplastics in Shanghai were analyzed, it has been claimed that suspended atmospheric microplastic particles were predominantly fibrous (67%) preceding fragments (30%) and granules (3%) [35].

While analyzing the microplastics in the coastal and marine environment, the predominance of fibers has been noted. The studies that extracted microplastic particles from the Yangtze Estuary and the East China Sea by floatation method have characterized the microplastic particles with the help of a stereomicroscope. They have differentiated the particles as fibers, films, granules, and spherules, out of which fiber is the most common category which accounts for 79.1% of total microplastic particles [17]. Similarly, Lusher et al. disclosed 95% contribution of microfibers in microplastic contamination in surface and sub-surface waters of Arctic Polar [56]. Microplastic particles in the coastal sands of Girgaon, Mumbai; Tuticorin; and Dhanushkodi were mostly fibrous in nature, whereas few were of granules and films [57]. In line with this, in the coastal region of Tamil Nadu (25 beaches were examined), the predominant form of microplastics was fragments and fibers [58]. The analysis of marine sediments along the Belgian coast by Claessens et al. has noted a similar trend of a higher contribution of fibers [59].

On the other hand, the analysis of farmland soils has also revealed a higher contribution of fibers among the different forms of microplastics (fragments, films, and pellet) found. A higher level of polypropylene is noticed which is then followed by polyethylene and polyester [28]. When the animal-based traditional medicinal materials were analyzed for microplastic contamination, a higher level of fibers (84.68%) was noted with a minimum contribution of fragments (15.32%) [41]. However, the gastrointestinal tracts of terrestrial birds were also found to be contaminated with microplastic fibers (87.72%) and fragments (12.28%). In addition to synthetic fibers, natural fibers were also observed in the intestinal tracts of birds [60]. Moreover, the analysis of table salt from 8 different countries for the microplastic contamination has mentioned that 25.60% of microplastics were of filament nature [45]. Figure 2 shows the contribution level of each form (fibers, fragments, granules, films, and pellets) of microplastics as per previous findings.

The researchers who have reported fibrous materials as high-level contributors also mentioned that these fibrous materials resemble the fibers being used in textile materials [61]. The fibers which drop out (microfiber shedding) from the textile materials during different phases can potentially contaminate the environment, which is proven by the studies that found plastic fibers in different environment.

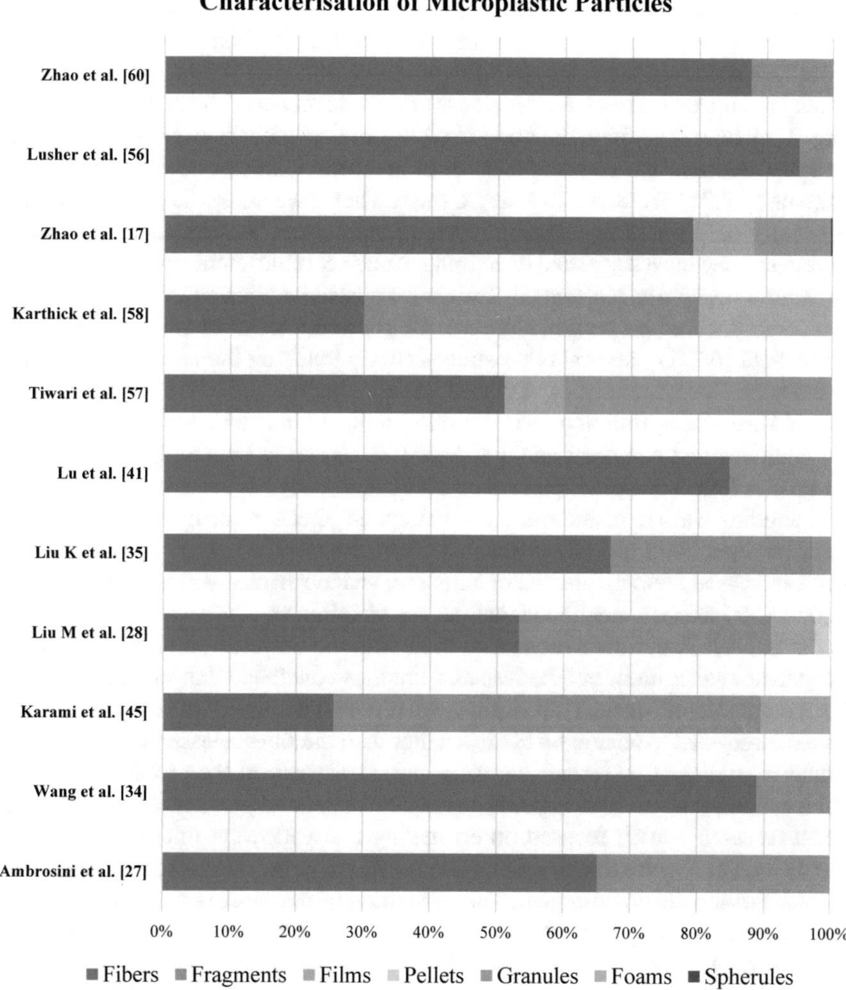

Fig. 2 Domination of microfibers in microplastic particles

5 Research Status of Microfiber Pollution

With microfibers as a major contributor to microfiber pollution, few studies back-tracked the source of microfiber and noted that domestic laundry is the potential source of leaching. As the public sewage treatment plants are not made to filter these tiny fibers, the leaching of microfiber to the sea or aquatic life system is unavoidable. The increased use of synthetic cloths is identified as the major source of microfiber pollution. The domestic laundry process is identified as the major source of microfiber release to the aquatic environment. The mechanical agitation and treatment of different chemicals like detergent, bleach, etc., are also proclaimed as the major source of microfiber generation from textile. Napper et al. studied the microfiber shedding behavior of different textile materials washed under different laundry parameters. They have detailed the quantification method of microfiber shedding. They have quantified the fibers shed in terms of count as well as weight. They have reported that the washing of 6 kg of synthetic textiles can shed fibers in the range of 1,37,951–7,28,789 in a single wash. They have noted the significant effect of laundry parameters like wash temperature and use of detergents and softeners. Moreover, they have suggested examining the effect of different washing machines, wash duration, spin speed, and other washing-related parameters [62]. As the impact of laundry detailed by preliminary research showed a lot of research requirements in this field [62, 63], several other studies concentrated on the laundry process and parameters. Reports analyzed different washing parameters like washing machine type [64], washing duration [65], washing temperature [66], water volume [67], and additives like detergent and softeners [68, 69] on the microfiber release from the textile. Francesca et al. performed a detailed study on these washing parameters including wash temperature, use of detergents, mechanical agitation in washing, water hardness, and use of detergents and softeners. The use of powdered detergents, increased temperature, water hardness, and mechanical agitation increase the shedding. It has been mentioned that the use of softeners while washing can reduce shedding [70]. Experimentations also tried with the recycled polyester material as a sustainable alteration, but the research findings concluded that the recycled fiber sheds more microfiber than the virgin synthetics and the size of the microfiber sheds from the recycled synthetic textile is smaller than the fiber released from the virgin synthetic textiles. This further improves the seriousness of the problem by stating that recycling is no more sustainable in the case of apparel products [71]. In the most recent research, which focused on estimating the worldwide microfiber emission into the aquatic environment, around 0.28 million tons of microfibers are estimated to enter the aquatic environment. The estimation of microfibers was done based on microfiber detachment rate, volume of laundry effluents, water treatment method, and type of washing machines used. With these parameters, they have estimated the region-wise contribution and claimed that Asia contributes around 65% of total microfibers across the world [72].

Various domestic laundry aids are also commercially available which can control the microfiber emission during domestic laundering. ‘Cora Ball’ is a laundry ball that is commercially available and can be used during laundry to catch the microfibers shed from the garments and prevents them from entering into the wash effluent. A researcher studied the efficiency of the commercially available ‘Cora Ball’ while washing 100% polyester fleece blanket. It has been concluded that Cora Balls have the potential to reduce the microfibers in the effluent up to 26% [73]. Another promising product that can reduce microfiber emissions during domestic laundry is LUV-R filters. These filters can be attached to the washing machine and they can prevent the fibers from entering the environment. These filters are capable of reducing fiber emissions up to 87%. The study which compared the performance of Cora Ball and LUV-R filter has concluded that in both weight and lengthwise reduction, the

LUV-R filter was noted with good performance [73]. The other commercially available products that help in the reduction of microfiber emission via laundering are ‘PlanetCare External filter,’ ‘Guppy friend bag,’ and ‘Filtrol 160™ external filter.’ The effectiveness of these products is also analyzed by the Swedish Environmental Protection Agency. The filters are capable of reducing the microfibers in the water approximately by 30–60%, and it is important to note that filters are capable of capturing only those fibers which are greater than 50 μm in size. However, fibers are even smaller than 50 μm and those cannot be filtered with the use of these filters. It has also been concluded that as of now, no solution can control 100% of microfiber emission in the wash effluent during domestic laundry [74].

6 Summary and Research Gap

Most of the researches were focused on microfiber shedding in the washing stage, and the possible solutions were made to control fiber release in the domestic laundry. The laundry aids (in-drum and external filters) are a short-term solution and not a proactive one. They are only preventing the fibers to enter the environment by capturing them after shed from the fabrics. However, they are not preventing the generation of fibers from the fabric surface. Moreover, the fibers which are captured using these filters and bags again end up in the environment as the proper way of disposal of those collected fibers is not clearly defined [75]. The main focus is given in the laundry process, whereas the shedding of fibers is possible in all the stages of the life cycle of the textile materials [55]. To overcome the problems with the available solutions, the majority of researchers have suggested more researches from the textile and fashion side [68]. The existing research works are done on the analysis of washing parameters, whereas the studies on the fiber and fabric properties like fiber type, yarn type, and fabric construction on the shedding are limited and more studies are essential to draw a proper conclusion on the shedding behavior of textile materials [76]. It has been suggested that the identification of textile parameters that accelerate the shedding can result in the key idea to reduce the shedding which could be a long-term solution for microfiber pollution [77]. As synthetic textiles are identified as the major sources of microplastic in the marine environment, it is highly necessary to initiate research from the textile design and manufacturing side to reduce the microfibers shedding [78].

References

1. Maiti R (2020) Fast fashion: its detrimental effect on the environment. https://earth.org/fast-fashions-detrimental-effect-on-the-environment/. Accessed on 15 May 2021
2. Ross CB (2017) What synthetic materials are doing to our environment. https://www.the-sustainable-fashion-collective.com/2017/04/11/synthetic-materials-environment. Accessed on 15

May 2021
3. Stone C, Windsor FM, Munday M, Durance I (2020) Natural or synthetic—how global trends in textile usage threaten freshwater environments. Sci Total Environ 718:134689. https://doi.org/10.1016/j.scitotenv.2019.134689
4. Hayes A (2021) Fast fashion. https://www.investopedia.com/terms/f/fast-fashion.asp. Accessed on 15 May 2021.
5. Fast Fashion's Detrimental Effect on the Environment (2020) by Editorial Board—Earth.org https://impakter.com/fast-fashion-effect-on-the-environment. Accessed on 15 May 2021
6. Brophy K (2020) Fast fashion 1—Why is the fashion industry an environmental problem? Inst Mol Sci Eng. https://wwwf.imperial.ac.uk/blog/molecular-science-engineering/2020/09/18/fast-fashion-1-why-is-the-fashion-industry-an-environmental-problem/. Accessed on 15 May 2021
7. Dardana C (2021) The destruction of unsold clothes shows the dark side of fast fashion. https://www.lifegate.com/fast-fashion-overproduction. Accessed on 15 May 2021
8. Sarah Young (2019) The real cost of your clothes: These are the fabrics with the best and worst environmental impact. https://www.independent.co.uk/life-style/fashion/fabrics-environment-fast-fashion-eco-friendly-pollution-waste-polyester-cotton-fur-recycle-a8963921.html. Accessed on 15 May 2021
9. Furstenberg D (2021) Natural vs. synthetic fiber: what's the difference? https://www.masterclass.com/articles/natural-vs-synthetic-fibers#advantages-of-using-synthetic-fibers. Accessed on 15 May 2021
10. The Dangers of Synthetic Fibers and Fabrics on the Environment (2018) https://superegoworld.com/blogs/the-world/the-dangers-of-synthetic-fibers-and-fabrics-on-the-environment. Accessed on 15 May 2021
11. Singh Z, Bhalla S (2017) Toxicity of synthetic fibres & health. Adv Res Text Eng 2(1):1012. https://austinpublishinggroup.com/textile-engineering/fulltext/arte-v2-id1012.php
12. Plastic Market Size, Share & Trends Analysis Report By Product (PE, PP, PU, PVC, PET, Polystyrene, ABS, PBT, PPO, Epoxy Polymers, LCP, PC, Polyamide), By Application, By End-use, By Region, and Segment Forecasts, 2021–2028 (2021) https://www.grandviewresearch.com/industry-analysis/global-plastics-market. Accessed on 15 May 2021
13. Ellen MacArthur Foundation (2017) A new textiles economy: redesigning fashion's future, http://www.ellenmacarthurfoundation.org/publications. Accessed on 15 Apr 2020
14. Plastic & Clothing (2019) https://www.theconsciouschallenge.org/ecologicalfootprintbibleoverview/plastic-clothing. Accessed on 15 May 2021
15. Bird S (2019) Synthetic fibers in clothing contribute to pollution. https://www.indianaenvironmentalreporter.org/posts/synthetic-fibers-in-clothing-contribute-to-pollution. Accessed on 15 May 2021
16. Ians (2019) Synthetic fibres contribute to plastic pollution. https://www.thehindu.com/sci-tech/energy-and-environment/synthetic-fibres-contribute-to-plastic-pollution/article26305420.ece. Accessed on 15 May 2021
17. Zhao S, Zhu L, Wang T, Li D (2014) Suspended microplastics in the surface water of Yangtze Estuary System, China: first observations on occurrence, distribution. Mar Pollut Bull 86(1–2):562–568. https://doi.org/10.1016/j.marpolbul.2014.06.032
18. Li C, Wang X, Liu K, Zhu L, Wei N, Zong C, Li D (2020) Pelagic microplastics in surface water of the Eastern Indian Ocean during monsoon transition period: abundance, distribution, and characteristics. Sci Total Environ 755:142629. https://doi.org/10.1016/j.scitotenv.2020.142629
19. Sanchez-Vidal A, Thompson RC, Canals M, de Haan WP (2018) The imprint of microfibres in southern European deep seas. *PLoS ONE* 13(11):e0207033. https://doi.org/10.1371/journal.pone.0207033
20. Absher TM, Ferreira SL, Kern Y, Ferreira AL, Christo SW, Ando RA. (2018) Incidence and identification of microfibers in ocean waters in Admiralty Bay, Antarctica. Environ Sci Pollut Res. https://doi.org/10.1007/s11356-018-3509-6
21. Castañeda RA, Avlijas S, Simard MA, Ricciardi A (2014) Microplastic pollution in St. Lawrence River sediments. Can J Fish Aquat Sci71(12):1761–1771. https://doi.org/10.1139/cjfas-2014-0281

22. Jiang C, Yin L, Li Z, Wen X, Luo X, Hu S, Yang H, Long Y, Deng B, Huang L, Liu Y (2019) Microplastic pollution in the rivers of the Tibet Plateau. Environ Pollut 249:91–98. https://doi.org/10.1016/j.envpol.2019.03.022
23. Mani T, Primpke S, Lorenz C, Gerdts G, Burkhardt-Holm P (2019) Microplastic pollution in benthic midstream sediments of the Rhine River. Environ Sci Technol 53(10):6053–6062. https://doi.org/10.1021/acs.est.9b01363
24. Eriksen M, Mason S, Wilson S, Box C, Zellers A, Edwards W, Farley H, Amato S (2013) Microplastic pollution in the surface waters of the Laurentian Great Lakes. Mar Pollut Bull 77(1–2):177–182. https://doi.org/10.1016/j.marpolbul.2013.10.007
25. Egessa R, Nankabirwa A, Ocaya H, Pabire WG (2020) Microplastic pollution in surface water of Lake Victoria. Sci Total Environ 741:140201. https://doi.org/10.1016/j.scitotenv.2020.140201
26. Sang W, Chen Z, Mei L, Hao S, Zhan C, Zhang W, Li M, Liu J (2021) The abundance and characteristics of microplastics in rainwater pipelines in Wuhan, China. Sci Total Environ 755(2):142606. https://doi.org/10.1016/j.scitotenv.2020.142606
27. Ambrosini R, Azzoni RS, Pittino F, Diolaiuti G, Franzetti A, Parolini M (2019) First evidence of microplastic contamination in the supraglacial debris of an alpine glacier. Environ Pollut. https://doi.org/10.1016/j.envpol.2019.07.005
28. Liu M, Lu S, Song Y, Lei L, Hu J, Lv W, Zhou W, Cao C, Shi H, Yang X, He D (2018) Microplastic and mesoplastic pollution in farmland soils in suburbs of Shanghai, China. Environ Pollut.https://doi.org/10.1016/j.envpol.2018.07.051
29. Chen Y, Leng Y, Liu X, Wang J (2019) Microplastic pollution in vegetable farmlands of suburb Wuhan, central China. Environ Pollut 257:113449. https://doi.org/10.1016/j.envpol.2019.113449
30. Huang Y, Liu Q, Jia W, Yan C, Wang J (2020) Agricultural plastic mulching as a source of microplastics in the terrestrial environment. Environ Pollut 260:114096. https://doi.org/10.1016/j.envpol.2020.114096
31. Sun Xiaoren (1991) A study on soil temperature when mulched by transparent plastic film. In: Energy conservation in buildings
32. Corradini G, Meza P, Eguiluz R, Casado F, Huerta-Lwanga E, Geissen V (2019) Evidence of microplastic accumulation in agricultural soils from sewage sludge disposal. Sci Total Environ 671:411–420. https://doi.org/10.1016/j.scitotenv.2019.03.368
33. Zhang J, Wang L, Halden RU, Kannan K (2019) Polyethylene terephthalate and polycarbonate microplastics in sewage sludge collected from the United States. Environ Sci Technol 6(11):650–655. https://doi.org/10.1021/acs.estlett.9b00601
34. Wang X, Li C, Liu K, Zhu L, Song Z, Li D (2019) Atmospheric microplastic over the South China Sea and East Indian Ocean: abundance, distribution and source. J Hazard Mater. https://doi.org/10.1016/j.jhazmat.2019.121846
35. Liu K, Wang X, Fang T, Xu P, Zhu L, Li D (2019) Source and potential risk assessment of suspended atmospheric microplastics in Shanghai. Sci Total Environ 675:462–471. https://doi.org/10.1016/j.scitotenv.2019.04.110
36. Gasperi J, Dris R, Mirande-Bred C, Mandin C, Langlois V, Tassin B (2015) First overview of microplastics in indoor and outdoor air. In 15th EuChemMS international conference on chemistry and the environment. https://hal.archives-ouvertes.fr/hal-01195546/
37. Kaya AT, Yurtsever M, Bayraktar SÇ (2018) Ubiquitous exposure to microfiber pollution in the air. Eur Phys J Plus 133:488. https://doi.org/10.1140/epjp/i2018-12372-7
38. Cauwenberghe LV, Claessens M, Vandegehuchte MB, Janssen CR (2015) Microplastics are taken up by mussels (*Mytilus edulis*) and lugworms (*Arenicola marina*) living in natural habitats. Environ Pollut 199:10–17. https://doi.org/10.1016/j.envpol.2015.01.008
39. Hossain SM, Sobhan F, Uddin MN, Sharifuzzaman SM, Chowdhury SR, Sarket S, Chowdhury MSN (2019) Microplastics in fishes from the Northern Bay of Bengal. Sci Total Environ 690:821–830. https://doi.org/10.1016/j.scitotenv.2019.07.065
40. Riberio F, Okoffo ED, O'Brien JW, Fraissinet-Tachet S, O'Brien S, Gallen M, Samanipour S, Kaserzon S, Muller JF, Galloway T, Thomas KV (2020) Quantitative analysis of selected plastics in high-commercial-value Australian seafood by pyrolysis gas chromatography mass spectrometry. Environ Sci Technol 54(15):9408–9417. https://doi.org/10.1021/acs.est.0c02337

41. Lu S, Qiu R, Hu J, Li X, Chen Y, Zhang X, Cao C, Shi H, Xie B, Wu W, He D (2019) Prevalence of microplastics in animal-based traditional medicinal materials: Widespread pollution in terrestrial environments. https://doi.org/10.1016/j.scitotenv.2019.136214
42. Selonen S, Dolar A, Kokalj AJ, Skalar T, Dolcet LP, Hurley R, van Gestel CAM (2019) Exploring the impacts of plastics in soil—The effects of polyester textile fibers on soil invertebrates. Sci Total Environ 400:134451. https://doi.org/10.1016/j.scitotenv.2019.134451
43. Jemec A, Horvat P, Kunej U, Bele M, Kržan A (2016) Uptake and effects of microplastic textile fibers on freshwater crustacean *Daphnia magna*. Environ Pollut 219:201–209. https://doi.org/10.1016/j.envpol.2016.10.037
44. Lee H, Kunz A, Shim WJ, Walther BA (2019) Microplastic contamination of table salts from Taiwan, including a global review. Sci Rep 9:10145. https://doi.org/10.1038/s41598-019-46417-z
45. Karami A, Golieskardi A, Choo CK, Larat V, Galloway TS, Salamatinia B (2017) The presence of microplastics in commercial salts from different countries. Sci Rep 7:46173. https://doi.org/10.1038/srep46173
46. Diaz-Basantes M, Conesa JA, Fullana A (2020) Microplastics in honey, beer, milk and refreshments in Ecuador as emerging contaminants. Sustainability 12:5514. https://doi.org/10.3390/su12145514
47. Muniasamy GK, Guevara FP, Martinez IE, Shruti VC (2020) Branded milks—Are they immune from microplastics contamination? Sci Total Environ 714:136823. https://doi.org/10.1016/j.scitotenv.2020.136823
48. Conti GO, Ferrante M, Banni M, Favara C, Nicolosi I, Cristaldi A, Fiore M, Zuccarello P (2020) Micro- and nono-plastics in Edible fruits and vegetables: the first diet risks assessment for the general population. Environ Res 187:109677. https://doi.org/10.1016/j.envres.2020.109677
49. Cox KD, Covernton GA, Davies HL, Dower JF, Juanes F, Dudas SE (2019) Human consumption of microplastics. Environ Sci Technol. https://doi.org/10.1021/acs.est.9b01517
50. Ragusa A, Svelato A, Santacroce C, Catalano P, Notarstefano V, Carnevali O, Pap F, Rongioletti MCA, Baiocco F, Draghi S, D'Amore E, Rinaldo D, Matta M, Giorgini E (2021) Environ Int 146:106274. https://doi.org/10.1016/j.envint.2020.106274
51. Abbasi S, Turner A (2021) Human exposure to microplastics: a study in Iran. J Hazard Mater 403:123999. https://doi.org/10.1016/j.jhazmat.2020.123799
52. Goodman KE, Hare JT, Khamis ZI, Hua T, Sang QA (2021) Exposure of human lung cells to polystyrene microplastics significantly retards cell proliferation and triggers morphological changes. Chem Res Toxicol 34(4):1069–1081. https://doi.org/10.1021/acs.chemrestox.0c00486
53. Boucher J, Friot D. Primary microplastics in the oceans: a global evaluation of sources. IUCN, Gland, Switzerland. https://doi.org/10.2305/IUCN.CH.2017.01
54. Rathinamoorthy R, Raja Balasaraswathi S (2020) A review of the current status of microfiber pollution research in textiles. International Journal of Clothing Science and Technology. https://doi.org/10.1108/IJCST-04-2020-0051
55. Rathinamoorthy R, Raja Balasaraswathi S (2021) Effect of textile parameters on microfiber shedding properties of textiles. In Muthu SS (ed) Microplastic pollution, sustainable textiles: production, processing, manufacturing & chemistry. https://doi.org/10.1007/978-981-16-0297-9_1
56. Lusher A, Tirelli V, O'Connor I et al (2015) Microplastics in Arctic polar waters: the first reported values of particles in surface and sub-surface samples. Sci Rep 5:14947. https://doi.org/10.1038/srep14947
57. Tiwari, M., Rathod TD, Ajmal PY, Bhangare RC, Sahu SK (2019) Distribution and characterization of microplastics in beach sand from three different Indian coastal environments. Mar Pollut Bull. https://doi.org/10.1016/j.marpolbul.2019.01.055
58. Karthik R, Robin RS, Purvaja R, Ganguly D, Anandavelu I, Raghuraman R, Hariharan G, Ramakrishna A, Ramesh R (2018) Microplastics along the beaches of southeast coast of India. Sci Total Environ. https://doi.org/10.1016/j.scitotenv.2018.07.242

59. Claessens M, De Meester S, Landuyt LV, De Clerck K, Janssen CR (2011) Occurrence and distribution of microplastics in marine sediments along the Belgian coast. Mar Pollut Bull 62(10):2199–2204. https://doi.org/10.1016/j.marpolbul.2011.06.030
60. Zhao S, Zhu L, Li D (2016) Microscopic anthropogenic litter in terrestrial birds from Shanghai, China: not only plastics but also natural fibers. Sci Total Environ 550:1110–1115. https://doi.org/10.1016/j.scitotenv.2016.01.112
61. Browne MA, Crump, P, Niven SJ, Teuten E, Tonkin A, Galloway T, Thompson R (2011) Accumulation of microplastic on shorelines worldwide: sources and sinks. Environ Sci Technol 45:9175–9179. https://doi.org/10.1021/es201811s
62. Napper IE, Thompson RC (2016) Release of synthetic microplastic plastic fibres from domestic washing machines: Effects of fabric type and washing conditions. Mar Pollut Bull 112(1–2):39–45. https://doi.org/10.1016/j.marpolbul.2016.09.025
63. Browne MA, Galloway T, Thompson R (2007) Microplastic—an emerging contaminant of potential concern? Integr Environ Assess Manag 3(4):559–561
64. Yang L, Qiao F, Lei K, Li H, Kang Y, Cui S, An L (2019) Microfiber release from different fabrics during washing. Environ Pollut 249:136–143. https://doi.org/10.1016/j.envpol.2019.03.011
65. Hernandez E, Nowack B, Mitrano DM (2017) Synthetic textiles as a source of microplastics from households: a mechanistic study to understand microfiber release during washing. Environ Sci Technol 51(12):7036–7046. https://doi.org/10.1021/acs.est.7b01750
66. Zambrano, MC, Pawlak JJ, Daystar J, Ankeny M, Cheng JJ, Venditti RA (2019) Microfibers generated from the laundering of cotton, rayon and polyester based fabrics and their aquatic biodegradation. Mar Pollut Bull 142:394–407. https://doi.org/10.1016/j.marpolbul.2019.02.062
67. Kelly M, Lant NJ, Kurr M, Grant Burgess J (2019) Importance of water-volume on the release of microplastic fibers from laundry. Environ Sci Technol 53:11735–11744. https://doi.org/10.1021/acs.est.9b03022
68. Claire OL (2018) Fashion and microplastic pollution, investigating microplastics from laundry. Ocean Remedy. https://cdn.shopify.com/s/files/1/0017/1412/6966/files/Fashion_and_Microplastics_Ocean_Remedy_2018.pdf. Accessed on 15 Feb 2021
69. Almroth BMC, Åström L, Roslund S, Petersson H, Johansson M, Persson N-K (2018) Quantifying shedding of synthetic fibers from textiles; a source of microplastics released into the environment. Environ Sci Pollut Res 25:1191–1199. https://doi.org/10.1007/s11356-017-0528-7
70. De Falco F, Gullo MP, Gentile G, Di Pace E, Cocca M, Gelabert L, Brouta-Agnesa M, Rovira A, Escudero R, Villalba R, Mossotti R, Montarsolo A, Gavignano S, Tonin C, Avella M (2018) Evaluation of microplastic release caused by textile washing processes of synthetic fabrics. Environ Pollut 236:619–925. https://doi.org/10.1016/j.envpol.2017.10.057
71. Özkan İ, Gündoğdu S (2020) Investigation on the microfiber release under controlled washings from the knitted fabrics produced by recycled and virgin polyester yarns. The Journal of The Textile Institute. https://doi.org/10.1080/00405000.2020.1741760
72. Belzagui F, Gutiérrez-Bouzán C, Álvarez-Sánchez A, Vilaseca M (2020) Textile microfibers reaching aquatic environments: a new estimation approach. Environ Pollut 265(Part B):114889. https://doi.org/10.1016/j.envpol.2020.114889
73. McIlwraitha HK, Lina J, Erdle LM, Mallos N, Diamond ML, Rochman CM (2019) Capturing microfibers—marketed technologies reduce microfiber emissions from washing machines. Mar Pollut Bull 139:40–45
74. Filters for washing machines Mitigation of microplastic pollution, EnviroPlanning, The Swedish Environmental Protection Agency, AB Lilla Bommen, Göteborg (2018). https://www.planetcare.org/images/documents/Swedish-EPA-filter-report-dec-2018.pdf. Accessed on 15 May 2021
75. Napper IE, Barrett AC, Thompson RC (2020) The efficiency of devices intended to reduce microfibre release during clothes washing. Sci Total Environ. https://doi.org/10.1016/j.scitotenv.2020.140412

76. Microfiber shedding—Topic FAQ, Outdoor Industry Association, European Outdoor group. https://static1.squarespace.com/static/5aaba1998f513028aeec604c/t/5db83b049c1c7a6e4cd66432/1572354825000/Microfiber+Shedding+FAQ+FINAL.pdf. Accessed on 15 May 2021
77. Marine pollution from microplastic fibers, Briefing by Fauna & Flora International January 2018 https://cms.fauna-flora.org/wp-content/uploads/2018/06/Marine-pollution-from-microplastic-fibres.pdf. Accessed on 15 May 2021
78. Boucher J, Friot D (2017) Primary microplastics in the oceans: a global evaluation of sources. IUCN, Gland, Switzerland. https://doi.org/10.2305/IUCN.CH.2017.01

Enabling Circular Fashion Through Product Life Extension

D. G. K. Dissanayake

Abstract The concept of circular economy is gaining traction as an alternative way of overcoming sustainability issues embedded in the linear fashion system. Circular economy business models are established based on the concept that the resources are kept on using for a longer period. Extending product life, a key strategy of circular economy, aims to keep the product in use to the highest extent as possible through design and operational practices. This chapter offers a comprehensive overview of strategies that enables extending the life of a fashion product. Three major strategies of (i) design for long life, (ii) product service systems that enable collaborative consumption through repairing, exchange, rental and leasing services, and (iii) refashioning models which are facilitated by the producer, or consumer-based Do-It-Yourself (DIY) methods are discussed in detail, together with their pros and cons. This analysis provides useful insights for the designers, consumers and businesses to support the transition towards sustainable and circular fashion.

Keywords Circular fashion · Product life extension · Sustainable fashion · Collaborative consumption · Product service systems

1 Introduction

Fashion industry is a highly globalized and complex industry. Meanwhile, it is economically important due to its large scale of production and consumption. However, this industry contributes to create both environmental and social impacts by consuming resources, lands, water, chemicals, polluting the environment and creating social inequity [18]. Current fashion system, which is driven by 'take-make-use-throwaway' scenario, believes the resources are infinite [31]. In this linear system, raw materials are processed into products, which are thrown away after use. This is also known as the material economy where the environment is damaged and polluted

D. G. K. Dissanayake (✉)
Department of Textile and Apparel Engineering, University of Moratuwa, Katubedda 10400, Sri Lanka
e-mail: geethadis@uom.lk

S. S. Muthu (ed.), *Sustainable Approaches in Textiles and Fashion*, Sustainable Textiles: Production, Processing, Manufacturing & Chemistry,
https://doi.org/10.1007/978-981-19-0530-8_2

by material extraction, material processing, consumption and waste dumping. This linear model has recently experienced not only the environmental pressure, but also an economic pressure due to rising prices of raw materials and depletion of resources.

Environmental issues occur at every stage of the fashion supply chain from fibre extraction up to the disposal of unwanted clothing. Natural fibres such as cotton are produced using massive areas of land, water and chemicals, whereas synthetic fibres such as polyester are extracted from fossil fuel [39]. Textile is categorized as the fourth highest impact category in terms of material consumption in EU after food, housing and transport [18]. Consumption of non-renewable resources in the textile industry is estimated as 98 million tons/year, and the input of fossil fuel feedstock is estimated to be reached 160 million tons by 2050 [19]. Total greenhouse gas emission of the textile industry was 1.2 billion tons of CO_2 equivalent in 2015 [41]. If the industry continues to operate in current status, it will intensify the resource scarcity and environmental pollution. Therefore, achieving sustainability in the textile industry is becoming increasingly important business strategy, yet remains challenging due to short production cycles and low utilization rates, which are mainly driven by fast fashion phenomenon. When the product life becomes shorter, energy, water and other resources used in the manufacturing process become wasted [43]. Extending product life is an alternative strategy to conserve resources and reduce waste generation.

Among all other stages in the textile supply chain, consumption and disposal practices of the consumers are very complex to understand and thus difficult to control. Frequent changes in fashion trends drive massive consumption and unacceptable disposal habits. Consumers do not pursue any emotional attachment or value for the goods that are fast produced and consumed goods, as they are meant to be disposable [13]. Quality of clothing is diminishing, and prices are falling, thus, clothing is increasingly becoming a throw away commodity. Global textile consumption reached 100 MT/year [54], and over 90 million clothes are end up in landfill [48]. Less than 1% of clothing is recycled back into clothing, and others are downcycled or thrown away. Around USD 100 billion worth of materials are lost every year due to landfilling or incineration of used clothing [19]. The amount of textile waste sent to landfill is continuing to grow as the consumption rises [9]. This planned obsolescence of premature clothing creates sustainability issues not only from the resource consumption viewpoint, but also due to the frequent environmental pollution created during production, consumption and disposal stages. These environmental issues are largely attributed to the consumer behaviour with regard to their consumption and disposal patterns [3]. Major sustainability initiatives taken by the fashion industry so far failed to address the consumption patterns as consumer interest on sustainability is not yet powerful enough to change their purchasing and consumption behaviours. To minimize the environmental issues associated with fast consumption and disposal habits, alternative ways of slowing down the purchasing and consumption patterns need to be emerged. Recent studies reported a notable shift towards these alternative consumption models, which are largely promoted by the concept of circular economy.

This book chapter has been arranged as follows. Section 2 briefly discusses the concept of circular economy, and Sect. 3 explains the relevance and importance of circular economy in the fashion industry. Section 4 comprises a detailed investigation on the product life extension strategies, which include design for longevity, product service systems and refashioning models. Section 5 highlights the key challenges for adopting product life extension models and Sect. 6 concludes the chapter.

2 Circular Economy

Circular economy gained traction during the last decade as a way of overcoming resource-intensive production and consumption models, while promoting the economic growth [26]. There are multiple definitions present regarding the circular economy concept, yet many of them explain somewhat similar idea, which emphasizes the importance of prioritizing the environment. Ellen MacArthur Foundation defines circular economy as a restorative or regenerative industrial system that shifts towards the use of renewable energy, eliminates toxic chemicals and waste, and aims for reusing of materials [20]. By analysing 114 definitions, Kirchherr et al. defined circular economy as 'an economic system that is based on business models, which replaces the 'end-of-life' concept with reducing, alternatively reusing, and recovering materials in production/distribution and consumption processes, thus operating at the micro level (products, companies, consumers), meso level (eco-industrial parks) and macro level (city, region, nation and beyond), with the aim to accomplish sustainable development, which implies creating environmental quality, economic prosperity and social equity, to the benefit of current and future generations' [33]. Bocken et al. [7] introduced three fundamental strategies of circular economy as follows:

(i) Slowing resource loops: this can be facilitated by designing the products for long use and also through product life extension by using service loops.
(ii) Closing the resource loops: this can be achieved through recycling, which closes the loop between post-use and production.
(iii) Narrowing the resource loop: this aims at using fewer resources than that are currently being used in the linear system.

Circular economy is often viewed as a waste management strategy [26], but circularity entails a broader life cycle approach than waste management. Circular economy aims at all the stages of product life cycle and considers alternative solutions to resource consumption including resource efficiency, better design and long use of products. Circular economy relies on three principles of (a) design out wastes, (b) keep products and materials in use and (c) regenerate natural systems [19]. However, most of these potentials of circular economy are less exploded. Implementing circular economy principles provides multiple benefits that represent three pillars of sustainability: economic, social and environmental. Economic benefits of circular models are mainly driven by the ability to restore the materials that are otherwise disposed in the linear system. Material circulation reduces the need of virgin material and resources and thus lessens the resource depletion and environmental pollution. Circular business models can open up new markets and revenue streams, while creating new employment opportunities. In Europe, 3.4 million people are already engaging in circular economy activities, which include repair, recycling, rental and leasing [42]. Transition from linear to circular economy requires collaboration of various stakeholders in the business, together with new policies, investments and vast amount of innovations to change the way that products are being produced and consumed.

3 Circular Fashion

The need of changing current production and consumption patterns in the fashion industry became apparent in recent years. Circular economy is considered as a promising approach to make this shift. In a circular economy, fashion and textiles are kept at their highest value during use phase and re-enter the material cycle after the use, without ending up as waste [19]. European commission identified textiles (fabrics and clothing) as a priority product category in circular economy [18]. Circular economy contrasts with the traditional linear system and focuses on continuous reuse of products, materials and renewable resources [7]. However, circular economy initiatives in fashion business are yet to emerge at large [11]. Sustainability principles are given the priority than circular economy principles, yet, the goals of sustainability are open-ended, whereas circular economy is viewed as closed-loop system [25]. The urgency in adopting circular economy principles into the fashion business is urged to reduce environmental pressure.

The term 'circular fashion' is defined as 'a fashion system that moves towards a regenerative model with an improved use of sustainable and renewable resources, reduction of non-renewable inputs, pollution and waste generation, while facilitating long product life and material circulation via sustainable fashion design strategies and effective reverse logistics processes' [15]. Instead of linear take-make-use-throwaway model, circular economy focuses on a 'take-make-use-recycle' scenario. Moving from linear to circular economy requires multiple business models, design strategies, methods, tools and most importantly, new ways of thinking [7]. The requirement to adopt more environmentally benign strategies in the fashion industry is driving the industry towards circular fashion business models [10]. In a circular fashion system, clothes are used more often, allowing to capture the full value of the product, and reuse or recycle them after the first life to recover the materials [19]. Dissanayake and Weerasinghe [15] highlight four key strategies that can make fashion circular.

(i) Resource efficiency: narrowing the resource loop by reducing resource consumption and waste generation, while using renewable and sustainable raw materials.
(ii) Circular design: design fashion clothing to be used for longer period and/or for several life cycles, which may include design for longevity, design for

customization, design for disassembly, design for recycling, and design for composting.

(iii) Product life extension: keep the product in use for the maximum possible duration by offering quality products and introducing product service systems such as repairing, rental and leasing programmes.

(iv) End-of-life circularity: at the end of first life cycle, clothing can be reused, recycled or remanufactured.

Circular approach to fashion aims to develop more sustainable fashion system by maintaining the value of product as long as possible and designing the products to support circularity [43]. It provides a life cycle approach, in which the use of sustainable and renewable resources is emphasized to support renewable and regenerative systems. Water, chemicals and fossil fuel consumption in fibre production should be minimized [39], and environmentally harmful fibres should be replaced with sustainable, degradable or organic raw materials. Bio-based and renewable inputs are essential in the textile production process. Reduction of resource consumption can be achieved through advanced technologies such as waterless dyeing, energy efficient machineries and eliminating waste generation [15]. Product design can support circularity by avoiding fibre blends and using mono materials to enhance recyclability [52]. Products can be designed to be suitable for several life cycles, which are meant to be circulated in a closed-loop system in terms of reuse, remanufacturing or recycling [43, 47]. Moreover, circular fashion focuses on extending the product life and enhances the utilization rates, as discussed below.

4 Product Life Extension

Current nature of the fashion industry encourages frequent purchases and premature obsolesce of clothing, resulting a substantial loss of material value. This extremely fast-moving industry needs to be slowed down to reduce the environmental burden and resource consumption. Previous studies demonstrated that greater environmental benefits can be achieved by extending the lifespan of clothing and reusing them without disposal [36, 44, 57]. The aim of product life extension is to keep the product in use for a maximum possible duration, thereby extracting highest possible value of the materials and resources embedded in the product. Designing high-quality clothing and keep them wearing more often would be effective in capturing value and reducing resource consumption and environmental pollution [19]. This can be viewed as extending resource loops where the use phase is stretched through long lasting and timeless design, and by encouraging maintenance and repair [24]. Keeping product in use gives stronger environmental benefits than recycling, in terms of material value and the amount of resources and energy required for recycling operations. The ultimate goal of extending the product life is to reduce the demand for new resources and maximize the value of consumed resources. Key strategies that enable product life extension are discussed in below sections.

4.1 Design for Longevity

Design stage accounts for 80% of the environmental impacts of a fashion product [21]. Design for long use can reduce the requirement of virgin materials, energy and other resources, while keeping the waste away from landfills. On the other hand, consumers are becoming more demanding for high-quality products and customization experience [23], which can be utilized positively to develop long-lasting products. Consumer satisfaction is the major factor that decides the useful life of a garment. One of the fundamental reasons for the consumers to discard clothing is that they no longer look good. To satisfy the consumer, product needs to fulfil both instrumental performance and psychological performance [44]. Instrumental performance refers to the facts such as fit, quality and durability, while beauty and fashion trend can be considered as psychological performance [44]. Both of these factors must be embedded during the product development process to develop a long-lasting product. Designers have the power to embed these factors into the product because they have the ability to influence on the selection of materials, silhouette, manufacturing quality and styling [10]. Design choices that are made in this stage significantly determine not only the product life, but also the environmental impacts associated with the product [39]. Without simply looking at fashion trends, designers must consider how products are made, used and disposed, together with their impacts to the environment and the society. Better design can reduce the material cost and disposal expenses, while minimizing the resource use [20]. Design for low impact materials and processes that minimize the discharge of harmful chemicals, reduce energy and water consumption, and support the recirculation of resources are integral parts of circular fashion. Design can be simplified to reduce material consumption, and silhouette can be selected to facilitate the extension of product life by recovering the material for reuse or recycling [10].

'Design for durability' refers to physical durability of the product, which facilitates long life without any failure [7]. Material selection plays an important part in this scenario, in which the selection of a durable material can minimize the rapid waste generation and extend the product life beyond one cycle through reuse or recycling. Product should be resilient to wear and tear and should withstand abrasion and washing [39]. For instance, Candiani Denim, a denim mill based in Milan, manufactures premium denim that is made to last and could be recycled at the end of life [11]. Other ways of increasing durability could include the use of strong seams in construction and application of long-lasting dyes and prints [19]. Offering a warranty at the point of sale is an effective way to demonstrate the retailer's commitment to durability [19]. Durable products can attract the loyalty of the customers and boost sales.

'Design for attachment and trust' refers to emotional durability, in which the product is loved and trusted longer [7]. Designers must ensure the product offers emotional connection to the user as long as possible [22]. Fast fashion culture fuelled by frequent seasonal changes compels consumers to dispose garments more often. Careful selection of colours and timeless designs can help consumer to accept that the product is not limited by seasonal fashion changes [39]. Timeless design is largely associated with its appearance. Timelessness can be achieved by designing fashionable products that do not go out of style quickly, simplifying the design or by incorporating historical context and craft practices [22]. Offering personalized products is another option to increase the emotional attachment to the product. In customizing process, customer becomes a co-designer by actively engaging in the process by providing personalized fit measurements and style alteration [14, 49]. This helps to improve customer's emotional attachment to the product, which enhances the product life in return. New technologies such as 3D body scanning and virtual prototyping are useful in product customization and better fit, while maintaining a low lead time. These digital technologies facilitate customer to analyse the styling, appearance and fit of the garment and suggest alterations, before the garment is being purchased [14].

Design for multi-style, adoptable and upgradable clothing can increase the product usability and delay the disposal [19]. Multi-style or convertible clothing means a particular garment can be worn in few different ways, for instance, a dress can be worn as a skirt by folding few lines. Convertible clothing is becoming more prominent with the increasing awareness through digital platforms. Various online tutorials, DIY videos and blogs are made available for consumers to demonstrate how a piece of clothing can be worn in few different ways. This concept is useful in reducing the number of purchases and increasing the utilization of clothes. Furthermore, garments can be designed to be adjusted for changing body sizes, which can delay the product disposal [15]. Russian label MUR makes adjustable clothing, in which garments are made for one size that can be adjusted for any body size using straps and ties.

4.2 *Product Service Systems (PSSs) and Collaborative Consumption*

Product service systems enable companies to sell the product together with the service. They contribute to the dematerialization of the economy by reducing material and energy consumption [6]. PSSs allow consumers to move from unsustainable consumption patterns to more sustainable consumption. Manzini et al. [40] defined a PSS as a 'business innovation strategy offering a marketable mix of products and services jointly capable of fulfilling a client's needs and/or wants – with higher added value and a smaller environmental impact as compared to an existing system or product. A major element of a PSS is that a consumer's need is met by selling utility instead of providing a product. In essence the right of product ownership is shifted from a client to the producer or service provider' (p. 30). PSSs decouple the economic success from material consumption and thus reduce the environmental impact associated with traditional retailing models [5]. PSSs can be categorized into three types: product-oriented PSS, use-oriented PSS and result-oriented PSS [56]. In product-oriented PSS, product is sold with a service contract, which may include free maintenance, repair, upgrading and substitution services over a defined period of time [40]. When the contract period is over, the producer may take back the product. Use-oriented PSS offers product sharing options without transferring the product ownership [1]. In result-oriented PSS, a result or competency is sold instead of a product [6].

PSSs are viewed as one of the major drivers of a circular economy. According to Brandstotter et al. [8], PSS tries to reach the goals of sustainable development in following ways.

Economic: economic advantages for the producer by shifting from production to service provision and for the user by only paying for consumed services.

Social: a wider group of people can access inexpensive services without the need of buying costly equipment. PSS also has potentials towards a bigger labour market as a service provider.

Environmental: by reusing and recycling products instead of wasting resources and energy for producing new ones, the environmental pollution can be reduced.

PSSs bring new level of relationship between the consumer and the retailer. Consumers are facilitated to move away from traditional product ownership and benefit from more 'flexibility', 'freedom' and 'accessibility' [40]. Meanwhile, PSSs enable fast replacement of fashion clothing without polluting the environment or creating demand for new products. On the other hand, a systematic shift to PSSs enables companies to gain profits while reducing the environmental issues created by the demand for new materials. Manzini et al. [40] explained the organizational benefits of PSSs as:

(i) New market development: product differentiation to offer of a new product service mix, or access markets that are unable to afford the costs of individual product ownership.
(ii) Increased flexibility: respond more rapidly to changing consumption market.
(iii) Cost reduction: reduction of prices of the product through higher systemic efficiencies.
(iv) Longer-term client relationships: new level of client relationships that result in longer-term profits.
(v) Improved corporate identity: to respond to the demands and be 'responsible and transparent'.
(vi) Improved market and strategic positioning: because of 'green consumerism' and existing and future environmental legal requirements/or restrictions, e.g. extended producer responsibility, resources taxes, environmental performance labelling and standards, and specific international agreements.

Collaborative consumption business models are enabled by means of PSSs and are based on the concept of sharing economy. Belk [4] explained the concept of sharing economy as the 'acquisition and distribution of a resource for a fee or non-monetary compensation'. In collaborative consumption, goods and/or services are shared by a group of people through swapping, trading, lending, borrowing, renting or leasing options [45, 50]. Herein, product ownership is not fully transferred to the consumer, instead, a temporary access is granted for the period of consumption. That means the product usage is emphasized rather than the ownership [37]. This makes a clear difference between the 'consumption' and 'use', where retailers keep the ownership and provide the customers with use function. The term collaborative consumption is used interchangeably with PSSs. However, Belk [4] argues that collaborative consumption does not involve a transfer of ownership, and therefore that excludes swapping models. In contrast, Park and Armstrong [45] explain that collaborative consumption can be described into two modes: utility-based non-ownership and redistributed ownership. Utility-based non-ownership includes rental and leasing schemes where product ownership is not transferred, and redistributed ownership means the transfer of ownership with the product through activities such as swapping, auction or resale [45].

Technology and digital platforms have enabled collaborative consumption moving from local communities to global market places [29]. Online rental provides useful and convenient access to fashion items with affordable prices [55]. Those digital platforms provide customer an opportunity of browsing wide range of options and choose the best for an affordable cost. Collaborative consumption also brings many sustainability benefits, such as maximizing the usage of a product and delaying disposal. Increase in usage can reduce the premature obsolescence. In customer point of view, sharing eliminates the product ownership cost of the consumer, where actual cost of the product is shared among users. This concept also provides access to luxury fashion and a range of product choices [20, 37]. Collaborative consumption is undoubtfully changing the way we consume fashion. Various forms of PSSs and collaborative consumption models are discussed below.

4.2.1 Repairing

Repairing is defined as a product-oriented PSS where functionality and durability of the product are guaranteed through maintenance and repair [2]. In order to reduce the level of purchase and disposal, consumers are required to maintain their clothes by upgrading or repairing them. Many consumers do not show an interest towards repairing their clothes, because purchasing a new item found to be cheaper than repairing. Consumers are more likely to repair clothes if they are expensive or valued in some way [35]. Apparel brands such as Patagonia, Eileen Fisher and Filson are promoting circular fashion by encouraging consumers to purchase high-quality clothing that can be repaired, rather than replaced [13].

Retailer-driven repairing services are emerging, where consumers can bring their clothes to the store and get them repaired, either free of charge or at a cost [15]. For instance, Nudie Jeans offers free repairs to all Nudie customers by maintaining repair shops within the store and also provides repair kits for their customers [12]. Patagonia facilitates repairing clothes either in-store or as DIY (Do-It-Yourself). Guidance for own clothing repair is provided through IFIXIT tutorials online, which includes step-by-step guide to repair various items [46]. Norwegian fashion company Livid Jeans offers first time repair service when purchasing a garment [34]. In this type of business model, customers expect the retailers to sell long-lasting products with access to repair. By providing repair services, retailers can retain customer loyalty and thereby boost sales [13], while in-store repairing service can open up new revenue stream and employment opportunities.

Other than retailer initiatives, independent repair shops are evolving, which facilitate cloth repair and organize various related events such as sewing skill workshops. For instance, the Renewal Workshop (USA) collects used clothing from various brands, repairs and resells them. Through repairing, the company diverted 20,000 pounds of clothing from U.S. landfills in 2016 alone [13]. Clothes Doctor (UK) offers full repairing service package to the customers including delivery and return facilities [15]. Furthermore, emergence of repair cafés is witnessed in recent years, which helps to boost the community interests on clothing repair [13]. The Repair Café Foundation in Netherlands sets up various cafés where people can gather and learn how to repair [17]. This type of community repair workshops provides an opportunity for consumers for social gathering and collaborative work, thus replacing the pleasure of shopping. Moreover, repairing own clothes can develop an emotional attachment to the product, thereby extending the lifespan of clothing. Such events also develop deep connections with the society and the environment, while providing knowledge exchange opportunities [13]. Gribblehirst Community Centre in Auckland conducts monthly repair workshops for public free of charge and invites professionally trained menders to assist the participants to repair their clothes [17].

4.2.2 Peer-to-Peer Exchange

Peer-to-peer exchange is not completely a new phenomenon. Exchanging clothes between family members and friends has long existed. As consumers became more aware of the sustainability issues associated with clothing disposal, this exchange model is expanded beyond family to community and strangers. Consumers tend to buy clothes but end up without wearing or one-off use due to various reasons such as fit issues, dislikes, changing body size and no longer fit for the purpose. Clothes exchange is found to be one of the best informal ways to get rid of such unwanted clothing without putting pressure on the environment.

Peer-to-peer exchange is enabled through cloth swapping or buy and sell options. This can be viewed as a collaborative consumption with redistributed ownership. Swapping means people meet in person or online to exchange their clothes, without making a monetary transaction [29, 45]. Unwanted, yet still usable garments are given out and consumers receive other fashion items in exchange. While it's a non-monetary exchange, a fee may be paid to access the event or as a membership fee of the facilitating company [29]. Unlike in rental or leasing models, product ownership is redistributed in swapping [37]. Moreover, instead of swapping, consumers can sell their unwanted clothing and buy another second-hand clothing that are sold by others in the same platform. This allows consumers to continuously upgrade their wardrobes through exchange and remove unwanted clothing. Through peer-to-peer exchange models, garments are kept on using for longer and multiple times without discarding them. This delays the purchase of new garments, as consumers get an opportunity to renew their wardrobes without a cost involvement. Clothes swapping events are heavily advertised in public places such as pubs and using social media and gaining an increasing consumer attraction. Apart from that, clothes swap 'apps' are becoming popular where consumers can download the app, upload the pictures of pieces they want to swap and pick items they would like to have. These apps also allow consumers to buy and sell used clothing for an affordable price. United Wardrobe, Vinted and Swancy are few famous example apps that facilitate clothes swap, buy or sell, and millions of customers are already using such app facilities for clothes exchange.

Retailers have shown an interest to organize cloth exchange events to provide an opportunity for their customers to exchange unwanted garments [37]. Swapping has been converted to a profit-oriented business by companies such as Clothing Exchange and Trend Sales by providing a platform for consumers to buy, sell or swap clothes though a subscription fee. Users can post items online and negotiate prices or swap by agreeing shipment terms [27]. Share Your Closet is another platform where consumers can borrow clothing free of charge, and the lender earns points which could then be used to borrow another item. Users can keep the garment until others want to borrow.

4.2.3 Rental and Leasing Services

Renting or leasing clothes instead of purchasing is a feasible option to slow down the resource loop as that reduces the requirement of virgin material and resources. This is known as use-oriented PSS where selling the use or availability of a product is not owned by the customer [2]. In operating rental or leasing services, fashion libraries are maintained instead of retail shops, where consumers can rent or lease fashion clothing from the library. Consumers can either perform short-term clothing rental by making a one-off payment, or use a subscription-based model to rent more than one garment during the subscription period [19]. Short-term rental model is suitable to meet short-term needs of the consumer, avoiding a permanent purchase for a one-off requirement. For instance, someone can rent an evening dress for a special occasion on short-term rental basics, as the consumer believes that it is wasteful to purchase a garment to wear only one time [45]. Houdini Sportswear offers high-quality performance sportswear for a week or for the weekend at a cost of 10–25% of its retail price, which is an attractive alternative for the customers who need quality clothing for an affordable price [19].

Subscription-based model allows customer to pay a flat fee for a pre-defined period and obtain a continuous service, which means the customers can frequently change their wardrobe while keeping up with latest fashion trends, yet without increasing the resource consumption [15]. This is an attractive and cost-effective method for consumers as an alternative to frequent purchase of new items and also to gain access to luxury items that they may not be able to afford otherwise [19, 45]. Rent the Runaway (RTR) is a popular rental service which allows consumers to rent luxury designer clothing at a flat monthly price. Multiple clothing items can be rented using this subscription-based model. RTR is committed to sustainability by introducing reusable bags for delivery and return of clothing, and a toxic-free cleaning process [51]. Dress You Can, an Italian fashion rental retailer based in Milan, offers online and offline rental services for women, which includes dresses, shoes and accessories for special occasions [11]. The company gives access to an exclusive wardrobe of designer clothing at an affordable price. This business model is supported by three suppliers, i.e. (i) the consumers supplying unwanted personal items and clothes; (ii) well-known fashion brands that supply vintage or seasonal clothes and (iii) young and emerging designers who provide their collections with lower fixed costs [11].

Rental or leasing business model does not transfer the ownership of the garment to the consumer; instead, the ownership stays with the service provider, who is responsible for maintaining the garments including repair and distribution [3]. The main difference between rental and leasing models is the time duration, where leasing services allow customers to keep clothing for a longer period of time than that of a rental service. For instance, MUD Jeans offer 'lease a jeans' option that includes repair services. Customer can lease the jeans for 12 months and decides to keep beyond the leased period, swap with a new pair of jeans or return at the end of lease period [19]. In rental or leasing models, producer remains as the owner of the product, and therefore, the producer has to ensure that the product lasts for a long life in order to rent or lease the product multiple times. This way, the producer postpones the disposal of existing products or manufacturing of a new product [40]. Therefore, comparing with a traditional product, those products which are made for PSS must have better product quality in terms of materials and construction, and a longer lifespan than their traditional counterparts. Returning the product after use requires new level of customer-retailor relationship based on trust. This new level of relationship can be used to improve the product and service by obtaining customer feedback at the point of return. Customers' comments on fit and quality can be fed back to the manufacturing process to improve durability and customer satisfaction, leading to a long-lasting relationship with customers [19].

Rental and leasing models are proven to be profitable for certain clothing segments, where high profit margins and lower rental prices can be achieved if product is rented several number of times [19]. Retailers can satisfy customer needs with a fewer number of garments, decreasing new product being manufactured and entering to the market [34]. These models allow consumers to access designer and luxury clothing for an affordable price, which may not otherwise accessible. Rental and leasing models can also eliminate the burden of ownership for the consumers and thus repair and maintenance work [37].

4.3 Refashioning Models

Refashioning is a strategy that disassembled used clothing, restyling and reassembling them to make new clothing. Refashioning unwanted clothes is useful in extending the product life beyond one cycle and to divert waste from landfilling.

Currently, around 1% of used clothes are recycled back into clothing, which implies a substantial loss of resources [19]. Increasing consumer awareness regarding environmental impacts of clothing disposal can help them to rethink and redesign their clothing rather than throwing them away due to fit issues or becoming out-of-trend. Refashioning garments are increasingly facilitated by the producer, independent designers and through Do-It-Yourself (DIY) models. Refashioning is found to be a better option to extend the product life and recovering the embedded value of the materials, comparing with other end-of-life options such as recycling [16].

Refashioning of used clothing is facilitated by various channels. Sustainable designers convert the concept of refashioning into a new circular business model, where used clothes are collected, redesigned and resold [16, 28]. User engagement in refashioning their own garments is increasing as they prefer to restyle their unused clothing and wear them again. Participatory redesign workshops are emerging, in which users can bring their unwanted clothes and redesign them with the support of experts. In participatory design workshops, users are guided and instructed by expert designers to redesign their garments. These workshops may provide both restyling knowledge and sewing skills, depending on the facilities available. Various online platforms are emerging recently that demonstrate consumers to refashion their clothes based on DIY model, which is also known as Design-It-Yourself [30]. A growing number of independent designers are actively engaging in these web platforms in educating general public to refashion their wardrobe. For instance, Scratch and Stitch is an online platform that provides DIY tutorials and inspires consumers to refashion various clothing items, while keeping up with latest fashion trends [53]. Elizabeth made this, Fashionista, Trash to Couture and Cotton and Curls are few more famous online blogs that publish DIY tutorials, with a growing number of followers. DIY videos and tutorials have also become famous through social media sites. Many inspiring videos, tutorials and blogs are posted in social media such as Facebook, Instagram, YouTube, and Pinterest that users can follow and get inspired to transform their unwanted clothes to new fashion items. This model can be further expanded by retailers providing DIY tutorials in their company websites including patterns of the original styles and refashioning options for some selected styles. Retailers can also team up with independent designers who already run online platforms and direct their consumer base to get inspired and convert their unwanted clothing into useful pieces, thereby extending the garment life beyond one cycle.

In the current linear system, garments are not designed to be refashioned or remanufacture; thus, introducing new design strategies that support refashioning garments at the end of first life cycle can be a new design innovation. Garments should be designed in a way that they can be disassembled easily, redesigned and reassembled. Design for disassembly is a strategy that is useful to incorporate in the design process for the garments that are expected to be refashioned. Garments should be simple in design that assembled minimum number of component parts with relatively large fabric pieces in each, in order to facilitate easy disassembly and redesign [16]. Moreover, selection of appropriate fabrics that can be used beyond one season is required to sustain this strategy.

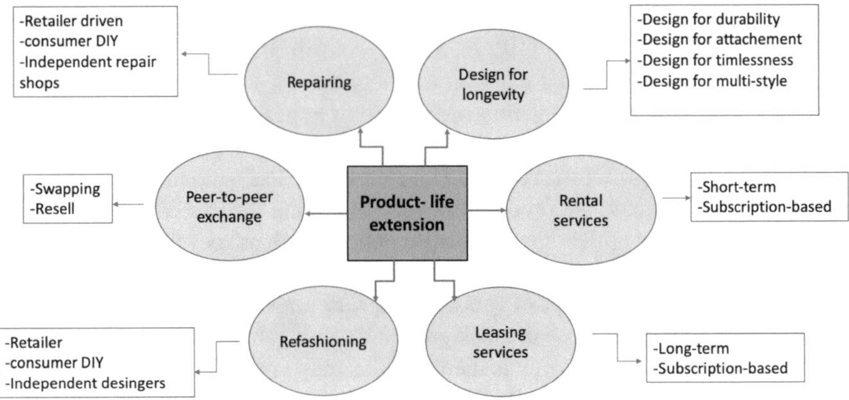

Fig. 1 Product life extension models

Figure 1 summarizes the types of product life extension models and their features, as discussed in this chapter.

5 Challenges for Adopting Product Life Extension Models

One of the main challenges contributing to negative perception of PSS models is the lack of trust in the provider [1]. Trust issues are closely associated with the quality, durability, hygiene issues and the other services provided. Lack of trust may hinder consumer interest on a particular PSS. For example, customer is unable to touch or feel the quality and durability of the product when using online rental services. This can lead to a performance risk where the product does not meet the consumer's functional expectation [38]. Customers' previous experience regarding the quality and durability of the products offered by the provider plays a vital role in selecting a product again from the same provider. Hygiene issues are associated with the customers' negative emotions towards an already used garments, which may include the factors such as how it has been used, cleaned and maintained [32]. Lack of awareness of the cleaning process may act as a barrier to accept PSS models [38]. In general, consumers are reluctant to exchange clothing, mainly due to hygiene issues. Sharing usually happens between family members or friends, rather than among strangers [4].

Another challenge is seen as the cultural shift required for the consumes 'having a need met as opposed to owing a product' [2]. Consumers may not be interested about ownerless consumption [2], and thus, convincing them to be a part of this model would be challenging. Marketing has a key role to play in this regard. For example, rental models should be attractive, fashionable and cost effective than the linear consumption model [19]. Significant challenges for rental models include lack of consumer awareness, social acceptance and inadequate regulations [55]. Starting from high-end segments would be useful for rental models as that can increase the brand value while developing consumer trust and relationship, which ultimately help to expand the model to mass-market segment [19]. However, consumers may rent or lease garments in order to enlarge their wardrobe, while practising the conventional purchasing patterns parallelly [12]. In that case, the environmental benefits may not be achieved as expected.

A PSS includes of tangible products and intangible services. Implementing PSSs is likely to be more complex for the businesses due to limited experience in terms of pricing, designing and delivering the PSS and also the associated risks [2]. This may also include lack of know-how in service design methods and management systems [40]. Effective logistics systems are required for delivery and return services, together with digital tracking systems to locate the product [19]. Operational cost for logistics management can act as a barrier towards the transition. Collaboration with existing delivery services could bring the cost down for the PSS provider. Apart from that, disruptive innovations such as clothes swapping can have negative impacts on business, as consumers may slow down the rate of purchasing high street fashion and consider exchanging their wardrobes [29]. On the other hand, uncertainty of quality and availability of sizes act as barriers for effective swapping [29].

PSSs have the potential to open up new markets and business opportunities. Retailer owned repair services can bring up a new business model, create employment opportunities and generate a revenue stream. However, the success of such PSS depends on the culture of the population [6]. Consumers in fast fashion culture are not emotionally attached to their clothes and hence do not motivate to extend their lives through repair [13]. They also prevent from repairing their clothes as they are not interested to wear the same garment for long time [1]. For the countries that cost of labour for repair or refashioning is more expensive than new clothes, operating such business models would be challenging [19]. Some retailers encourage consumers to repair their own clothing and provide facilities such as repair kits. However, previous studies reported that lack of repair skills and the time availability may discourage consumers to undertake repair services [13, 35]. Apart from that, clothing repair is considered as women's work and a sign of poverty [13].

Refashioning may not be preferred by some consumers due to the 'patchwork like' look in the refashioned garment [1]. Refashioning also needs time, effort and skill of the consumer, together with an emotional attachment to the product. Moreover, redesigning a garment is largely restricted by its original design, construction and the availability of usable area of fabrics [16]. Consumers find this as a difficult task, which needs creativity and time. They may prefer to get the pleasure of a shopping experience and the ownership of a new product, rather than working with an old piece of clothing. Apart from that, fashion consumption is mostly driven by the psychological needs of the consumers such as identity, personality, wealth and the social status. Extending the life of same product would not fulfil their emotional requirements. In that context, consumers may tend to ignore the quality and durability of products and select fast replacement option to fulfil their emotional needs.

Table 1 summarizes the key challenges present in each of the product life extension strategy, in consumer point of view.

Table 1 Challenges for using product life extension strategies

Product life extension strategy	Challenges
Design for longevity	– consumer interest on fast fashion – frequent changes in fashion trends – psychological needs of the consumer
Repair	– attachment to the product – repair skills – time and facilities needed
Rental/leasing	– hygiene issues – trust of the provider – quality and durability issues – absence of physical inspection in online rental schemes
Swapping	– hygiene issues – quality and durability issues – lack of trust on the previous owner
Refashioning	– time, skill, and facilities needed – lack of emotional attachment to the product – quality of the garment – design limitations – psychological needs of the consumer

In order to overcome challenges faced by the product life extension models, consumer awareness regarding the environmental benefit of such systems must be raised. Previous studies suggest that eco-minded consumers are willing to pay higher prices than the market average for a sustainable product. However, product life extension models are cheaper to use than purchasing a new garment, which could ideally attract a wider consumer base. Product cost is shared in collaborative consumption, and by using design strategies such as longevity and refashioning, product can be used for a longer and multiple time. These types of consumer cost saving strategies should be backed by heavy marketing campaigns, emphasizing their environmental credibility. Changing consumer mindset is the promising approach towards the success of product life extension models.

6 Conclusion

This study offers a holistic examination of the product life extension strategies that move fashion industry towards circular economy. It is worth understanding that collaborative fashion consumption is evolving, and more business opportunities and revenue streams are generating around that. Collaborative consumption increases the product utilization and brings sustainability benefits by reducing the requirement for new materials and delaying the waste generation. That also helps shifting consumers from frequent buying habits to alternative methods of consumption. The benefits of a circular fashion system create a win–win situation for both the retailers and consumers, yet significant challenges remain in adopting circular business models. New approaches should be developed to overcome barriers to operationalize PSSs, support collaborative consumption and other product-life extension scenarios such as repair and refashioning. For the successful implementation of PSSs, design for circularity must be embedded in the firm's business strategy. Design method enables product to have durability, timelessness, personalization, repairing and refashioning capabilities. Customer satisfaction is critical to sustain PSSs, and therefore, a successful PSS must incorporate customer feedback loops and continuous improvement strategies. Ultimately, circular economy encourages consumers to buy less, but better quality and increase the utilization rates. It also shifts the societal thinking on how clothes are bought and consumed.

References

1. Armstrong CM, Niinimäki K, Kujala S, Karell E, Lang C (2015) Sustainable product-service systems for clothing: exploring consumer perceptions of consumption alternatives in Finland. J Clean Prod 97:30–39. https://doi.org/10.1016/j.jclepro.2014.01.046
2. Baines TS, Lightfoot HW, Evans S, Neely A, Greenough R, Peppard J, Roy R, Shehab E, Braganza A, Tiwari A, Alcock JR, Angus JP, Basti M, Cousens A, Irving P, Johnson M, Kingston J, Lockett H, Martinez V, Michele P, Tranfield D, Walton IM, Wilson H (2007) State-of-the-art in product-service systems. Proc Inst Mech Eng Part B J Eng Manuf 221:1543–1552. https://doi.org/10.1243/09544054JEM858
3. Becker-Leifhold CV (2018) The role of values in collaborative fashion consumption—a critical investigation through the lenses of the theory of planned behavior. J Clean Prod 199:781–791. https://doi.org/10.1016/j.jclepro.2018.06.296
4. Belk R (2014) You are what you can access: sharing and collaborative consumption online. J Bus Res 67:1595–1600. https://doi.org/10.1016/j.jbusres.2013.10.001
5. Bell S, Davis B, Javaid A, Essadiqi E (2006) Final report on design of recyclable products. https://doi.org/10.13140/RG.2.2.18847.36009
6. Beuren FH, Gomes Ferreira MG, Cauchick Miguel PA (2013) Product-service systems: a literature review on integrated products and services. J Clean Prod 47:222–231. https://doi.org/10.1016/j.jclepro.2012.12.028
7. Bocken NMP, de Pauw I, Bakker C, van der Grinten B (2016) Product design and business model strategies for a circular economy. J Ind Prod Eng 33:308–320. https://doi.org/10.1080/21681015.2016.1172124
8. Brandstotter M, Haberl M, Knoth R, Kopacek B, Kopacek P (2003) IT on demand—towards an environmental conscious service system for Vienna (AT), 799–802. https://doi.org/10.1109/vetecf.2003.239737
9. Bukhari MA, Carrasco-Gallego R, Ponce-Cueto E (2018) Developing a national programme for textiles and clothing recovery. Waste Manag Res 36:321–331. https://doi.org/10.1177/0734242X18759190

10. Claxton S, Kent A (2020) The management of sustainable fashion design strategies: an analysis of the designer's role. J Clean Prod 268:122112. https://doi.org/10.1016/j.jclepro.2020.122112
11. Colucci M, Vecchi A (2020) Close the loop: evidence on the implementation of the circular economy from the Italian fashion industry. Bus Strateg Environ, 1–18. https://doi.org/10.1002/bse.2658
12. Corvellec H, Stål HI (2017) Evidencing the waste effect of Product-Service Systems (PSSs). J Clean Prod 145:14–24. https://doi.org/10.1016/j.jclepro.2017.01.033
13. Diddi S, Yan R-N (2019) Consumer perceptions related to clothing repair and community mending events: a circular economy perspective. Sustainability 11. https://doi.org/10.3390/su11195306
14. Dissanayake DGK (2020) Does mass customization enable sustainability in the fashion industry? In Fashion industry-an itinerary between feelings and technology. https://doi.org/10.5772/intechopen.88281
15. Dissanayake DGK, Weerasinghe D (2021) Towards circular economy in fashion: review of strategies, barriers and enablers. Circ Econ Sustain. https://doi.org/10.1007/s43615-021-00090-5
16. Dissanayake G, Sinha P (2015) An examination of the product development process for fashion remanufacturing. Resour Conserv Recycl 104:94–102. https://doi.org/10.1016/j.resconrec.2015.09.008
17. Durrani M (2018) "People gather for stranger things, so why not this?" Learning sustainable sensibilities through communal garment-mending practices. Sustainability 10. https://doi.org/10.3390/su10072218
18. EEA (2019) Briefing-textiles-in-europe-s-circular-economy
19. Ellen MacArthur Foundation (2017) A new textiles economy: redesigning fashion's future
20. Ellen MacArthur Foundation (2012) Towards the circular economy
21. European Commission (2014) Ecodesign your future: how ecodesign can help the environment by making products smarter, 1–12
22. Flood Heaton R, McDonagh D (2017) Can timelessness through prototypicality support sustainability? A strategy for product designers. Des J 20:S110–S121. https://doi.org/10.1080/14606925.2017.1352671
23. Gazzola P, Pavione E, Pezzetti R, Grechi D (2020) Trends in the fashion industry. The perception of sustainability and circular economy: a gender/generation quantitative approach. Sustainability 12:1–19. https://doi.org/10.3390/su12072809
24. Geissdoerfer M, Pieroni MPP, Pigosso DCA, Soufani K (2020) Circular business models: a review. J Clean Prod 277:123741. https://doi.org/10.1016/j.jclepro.2020.123741
25. Geissdoerfer M, Savaget P, Bocken NMP, Hultink EJ (2017) The circular economy—a new sustainability paradigm? J Clean Prod 143:757–768. https://doi.org/10.1016/j.jclepro.2016.12.048
26. Ghisellini P, Cialani C, Ulgiati S (2016) A review on circular economy: the expected transition to a balanced interplay of environmental and economic systems. J Clean Prod 114:11–32. https://doi.org/10.1016/j.jclepro.2015.09.007
27. Guldmann E (2016) Best practice examples of circular business models. Danish Environ Prot Agency. https://doi.org/10.13140/RG.2.2.33980.95360
28. Han SLC, Chan PYL, Venkatraman P, Apeagyei P, Cassidy T, Tyler DJ (2017) Standard vs. upcycled fashion design and production. Fash Pract 9:69–94. https://doi.org/10.1080/17569370.2016.1227146
29. Henninger CE, Bürklin N, Niinimäki K (2019) The clothes swapping phenomenon—when consumers become suppliers. J Fash Mark Manag 23:327–344. https://doi.org/10.1108/JFMM-04-2018-0057
30. Kejing L, Qi Z (2011) Green fashion design under the Concept of DIY. J Innov Sustain RISUS ISSN 2179-3565 2, 81. https://doi.org/10.24212/2179-3565.2011v2i1p81-86
31. Ki CW, Chong SM, Ha-Brookshire JE (2020) How fashion can achieve sustainable development through a circular economy and stakeholder engagement: a systematic literature review. Corp Soc Responsib Environ Manag 27:2401–2424. https://doi.org/10.1002/csr.1970

32. Kim NL, Jin BE (2021) Addressing the contamination issue in collaborative consumption of fashion: does ownership type of shared goods matter? J Fash Mark Manag 25:242–256. https://doi.org/10.1108/JFMM-11-2019-0265
33. Kirchherr J, Reike D, Hekkert M (2017) Conceptualizing the circular economy: an analysis of 114 definitions. Resour Conserv Recycl 127:221–232. https://doi.org/10.1016/j.resconrec.2017.09.005
34. Kongelf I, Camacho-Otero J (2020) Service design and circular economy in the fashion industry. In NordDesign 2020. https://doi.org/10.35199/norddesign2020.53
35. Laitala K, Klepp IG (2018) Care and production of clothing in Norwegian Homes: environmental implications of mending and making practices. Sustainability 10. https://doi.org/10.3390/su10082899
36. Laitala K, Klepp IG (2017) Clothing reuse: the potential in informal exchange. Cloth Cult 4:61–77. https://doi.org/10.1386/cc.4.1.61_1
37. Lang C, Joyner Armstrong CM (2018) Collaborative consumption: the influence of fashion leadership, need for uniqueness, and materialism on female consumers' adoption of clothing renting and swapping. Sustain Prod Consum 13:37–47. https://doi.org/10.1016/j.spc.2017.11.005
38. Lang C, Seo S, Liu C (2019) Motivations and obstacles for fashion renting: a cross-cultural comparison. J Fash Mark Manag 23:519–536. https://doi.org/10.1108/JFMM-05-2019-0106
39. Manshoven S, Chistis M, Vercalsteren A, Arnold M, Nicolau M, Lafond E, Fogh L, Coscieme L (2019) Textiles and the environment in a circular economy. Eur Top Cent Waste Mater Green Econ, 1–60
40. Manzini E, Vezzoli C, Clark G (2001) Product-service systems: using an existing concept as a new approach to sustainability. J Des Res 1. https://doi.org/10.1504/jdr.2001.009811
41. McKinsey&Company (2016) Style that's sustainable: a new fast-fashion formula
42. Mitchell P, James K (2015) Economic growth potential of more circular economies 28
43. Niinimäki K (2018) Sustainable fashion in a circular economy, sustainable fashion in a circular economy
44. Niinimäki K (2017) Fashion in a circular economy. In Henninger CE, Ryding D, Alevizou PJ, Goworek H (eds) Sustainability in fashion: a cradle to upcycle approach, 151–169. https://doi.org/10.1007/978-3-319-51253-2
45. Park H, Armstrong CMJ (2017) Collaborative apparel consumption in the digital sharing economy: an agenda for academic inquiry. Int J Consum Stud 41:465–474. https://doi.org/10.1111/ijcs.12354
46. Patagonia (2021) Product Care & Repair [WWW Document]
47. Pedersen ERG, Earley R, Andersen KR (2019) From singular to plural: exploring organisational complexities and circular business model design. J Fash Mark Manag 23:308–326. https://doi.org/10.1108/JFMM-04-2018-0062
48. Pedersen ERG, Gwozdz W, Hvass KK (2018) Exploring the relationship between business model innovation, corporate sustainability, and organisational values within the fashion industry. J Bus Ethics 149:267–284. https://doi.org/10.1007/s10551-016-3044-7
49. Piller FT, Müller M (2004) A new marketing approach to mass customisation. Int J Comput Integr Manuf 17:583–593. https://doi.org/10.1080/09511920420002731 40
50. Piscicelli L, Cooper T, Fisher T (2015) The role of values in collaborative consumption: insights from a product-service system for lending and borrowing in the UK. J Clean Prod 97:21–29. https://doi.org/10.1016/j.jclepro.2014.07.032
51. Rent the Runway (2020) A consumer-centered approach for managing post-consumer textile flows [WWW Document]
52. Sandvik IM, Stubbs W (2019) Circular fashion supply chain through textile-to-textile recycling. J Fash Mark Manag 23:366–381. https://doi.org/10.1108/JFMM-04-2018-0058
53. Scratch and Stitch (2021) 30 clothing refashion projects perfect for spring [WWW Document]
54. Shirvanimoghaddam K, Motamed B, Ramakrishna S, Naebe M (2020) Death by waste: fashion and textile circular economy case. Sci Total Environ 718:137317. https://doi.org/10.1016/j.scitotenv.2020.137317

55. Shrivastava A, Jain G, Kamble SS, Belhadi A (2021) Sustainability through online renting clothing: circular fashion fueled by Instagram micro-celebrities. J Clean Prod 278:123772. https://doi.org/10.1016/j.jclepro.2020.123772
56. Tukker A (2004) Eight types of product-service system: eight ways to sustainability? Experiences from suspronet. Bus Strateg Environ 13:246–260. https://doi.org/10.1002/bse.414
57. Woolridge AC, Ward GD, Phillips PS, Collins M, Gandy S (2006) Life cycle assessment for reuse/recycling of donated waste textiles compared to use of virgin material: an UK energy saving perspective. Resour Conserv Recycl 46:94–103. https://doi.org/10.1016/j.resconrec.2005.06.006

Challenges and Opportunities for Circular Fashion in India

Minakshi Jain

Abstract Circular fashion is a budding phenomenon in contemporary Indian fashion landscape, though the concept is antecedently woven in our traditional social fabric. To save our mother planet and its environment, as well as to minimize the depletion of natural resources from the adverse impact of fast fashion and consumerism circularization of production and utilization processes is of utmost importance. Adoption of circular fashion practices can be established as key to the socio-ecological-economic issues being faced by fashion manufacturers and consumers. It requires amalgamation of sustainable fashion technology with traditional textile crafts in all stages of manufacturing and use of textile products so that maximum utilization of waste generated from one segment can be done as raw material for the another one. Many innovative processes and techniques are being adopted by various Indian fashion houses across the supply chain as circular business models. It facilitates minimal consumption of virgin natural resources in addition to economic bonanza for the fashion industry troubled with financial crisis due to COVID-19.

Circular fashion is in its nascent stage in the modern Indian fashion industry confronting innumerable problems for adoption and recognition it ought to have. This paper deals with the challenges that come across during selection of raw material, dry and wet processing, cutting and stitching, packaging, transportation, retailing, use and post-use for circular fashion in India. Brilliant waves of opportunities surging to trigger circular fashion innovations and practices consequential to its perspective growth in local, regional and international market have also been discussed in the paper.

Keywords Circular fashion · Innovations · Recycling · Reuse · Supply chain · Sustainability

---

M. Jain (✉)
Government Girls College, Chomu, Rajasthan, India

S. S. Muthu (ed.), *Sustainable Approaches in Textiles and Fashion*, Sustainable Textiles: Production, Processing, Manufacturing & Chemistry,
https://doi.org/10.1007/978-981-19-0530-8_3

1 Introduction

1.1 Need of Circular Fashion

The fashion and textile industry is among one of the largest sectors that consumes a huge amount of natural resources that cause for environmental imbalance. The existing linear fashion model of buy, use and dispose is significantly accountable for negative impact of textile industry as well as economic breakdown. Fast fashion and impulsive buying outburst into abundance of low-quality non-degradable textile waste. The desire for spanking new designs in garments brings to wardrobe full of partially used and unused garments that mostly goes to landfills or for incineration.

Dry and wet processing involved in textile manufacture and fashion clothing production consumes huge amount of water, energy and other virgin resources from nature. Global industrial water pollution caused by textile processing is 20% [17]. The linear system of fashion industry requires non-renewable resources like oil and other chemicals to produce synthetic fibres in addition to a massive amount of water and chemical fertilizers for cultivation of natural fibres. Dyeing, finishing and other processing employ considerable quantity of toxic chemicals. More than 70 million trees are destroyed to make synthetic fibres like rayon and viscose and 70 million barrel oil is exhausted to meet the requirement of the fibres commonly used for apparel production worldwide. These synthetic fibres such as nylon and polyester can take centuries to get decomposed. Many harmful gases like nitrous oxide are also omitted during the production of synthetic fibres which induce global warming 300 times faster as compared to carbon dioxide.

The textile and fashion products are mostly produced in underdeveloped countries and remarkable environmental pollution has been observed in these countries which includes loss of natural resources, scarcity of clean water and energy and degradation of fertile soil. Besides pre-consumer and post-consumer wastes are not treated properly—neither recycled nor reused. A significant amount of hazardous waste are created due to ignorance that eventually transforms into toxins harmful for living beings as well as environment.

The linear system of fashion consumption includes designing, manufacturing, sales and disposal of fashion products [20]. By following this system, vast amount of precious raw material and natural resources is being wasted, especially if the lifespan of any article is very short. According to WRAP (2012) [26], around one-third of the fabric is wasted during production phase that is pre-consumer waste. Almost 80% of the textile fashion products turn into waste and thrown away within 6 months of their production [6]. Many of the garments from impulsive purchased are never worn [21]. According to Fletcher (2008) [15], up to 70% of our wardrobe contents remain unused. The number of wearing per garment has been decreased by 36% as compared to the number of wearing 15 years ago [11]. Besides, approximately 20% of the garments do not come to the market and remain unsold [23] which further increase the problem of textile waste production.

When we compare with other developed countries, the textile waste generation of Indian population is quite low. Indian traditional textile practices such as reuse, mending, renovation, multiple use, multi personal use, zero-waste designs of garments (as in saree, dhoti, lungi, etc.) and less emphasis on varied fashionable clothing are such factors that contribute to minimum wastage of textile. It is Indian inheritance of lifestyle imbibing use and reuse of garments among members of the family and even circulate them between relatives. These kind of customs, lifestyle and behaviour of sharing clothing with each other not only satisfy the crave for the search of new dresses and designs but also minimize the early disposal of any garment and ultimately turns into significant reduction in the textile waste of the country. These traditions corroborate the principle of circular economy for clothing and home textiles in Indian households and extensively contribute and motivate the populace towards circularity.

The currently prevailing linear system has geared up fast fashion that requires fast manufacturing, transportation and consumption; less use, reduced wearing and early disposal. It increases the actual cost of production, not only in financial terms but also in respect of exhaustion of precious natural resources resulting into significantly high impact of linear fashion on the environment. During the use of garment, maintenance and laundering further add to water pollution. Therefore the linear fashion consumption model is an oversaturated and oversized fashion system with big environmental impact [4].

Hence there is a need to create a better system for optimum use of resources and reduction of environmental pollution—a system that can make use of resources, processes and utilization pattern efficiently and intelligently. The linear system needs to be transformed into a closed-loop system in which every step of the supply chain as well as consumption pattern is turned into circular economic fashion model to attain the sustainable goals.

During the last few years circular fashion has been quite extensively encouraged across the world. It is based on reuse of raw natural resources, renewable energy and water; complete removal of the use of toxic chemicals and exterminate waste through thoughtful designing. Alternatively, it can be stated that in circular economy fashion products are supposed to be designed with ethical values in mind. These are to be produced with high longevity, with efficient use of resources and non-toxic material; they should be biodegradable, recyclable and reusable. For circular fashion more emphasis should be given to local products made of locally available raw material utilizing local resources and local knowledge. The raw material should be non-toxic, biodegradable, renewable and recyclable; the production process needs to be well-efficient, safe and go in line with ethical practices. On the consumers' side, the products should be used for the longest possible time, may be through good care, proper maintenance, mending, refurbishing, repairing and sharing among multiple users during their lifespan—second hand or swapping services can be used for this. When the product comes to the end of its one life it can be redesigned to give a new life to its components and reusable parts—components can be separated and should be recycled and reused for the creation of new products according to their nature. If the components are not suitable for recycling, their biological part can be used as compost to nourish the plants and other living creatures of the earth. In general, the life cycle of circular fashion products doesn't harm the environment and socio-economic constituents of the society. Rather it brings the human society towards their well-being, socio-economic upliftment of the country and the planet in wider perspective.

All the materials and processes used to make circular fashion products are essentially safe for the environment, effective in use, prepared and disposed in justifiable manner. These do not produce any kind of toxic or non-toxic waste—the waste from one stage of production or utilization is employed as the raw material, nutrient or resource for the next stages of production or consumption. New and virgin raw materials are used minimal and priority has been given to make use of alternative energy resources and recycled water to minimize the environmental impact of fashion production and consumption. Thus, these products essentially circulate among the users present in the society for longest possible duration.

India is a country where conservation and saving are more appreciated then consumption and expenditure. In today's fashion world, India is playing dual role both as producer and as consumer. The rising income of Indian community is transforming India from solely a manufacturing hub towards the key consumption state; therefore there are obvious identifiable challenges as well as opportunities emerging within India towards sustainability and circularity analogous to those faced by consumer-driven economies. The stakeholders have a major responsibility in converting the processes of fashion production from linear to circular, providing key solutions towards circular fashion production and services. Another most critical and important accountability is on the consumer side to adopt circular fashion and sustainable conduct in their consumption pattern with increased awareness and consciousness regarding circular fashion in order to ensure sustainability of resources available on the planet. Circular innovations in India are going in tandem with those in the Global landscape, taking into consideration the local conditions, unique lifestyle and clothing habits of Indian populace. Challenges and opportunities are being explored in all segments from branded machine-made garments to conventional locally hand-produced clothing. Incorporation of circular processes is driving the fashion industry towards sustainability ensuing increasing efficiency of both the local and global producers in addition to enable the shift towards minimizing the waste in all sectors of the supply chain.

The problem of waste generation rose from inconsiderate and excessive production and consumption pattern has given the approach to the concept of circular economy. Circular fashion is gaining impetus and increasing attention as a solution to overcome the problems arising from resource depletion and waste generation. Circular fashion is a system of fashion production that incorporates invigorating, regenerative and reproductive design and production techniques. Although the supply chain is the pivot point to take action towards circular fashion, all the stakeholders have their own roles and responsibilities for attaining the goal. At present, circular fashion is in its infancy and very less literature is available regarding this concept but this chapter is an effort to provide valuable insights concerning supply chain management indispensable for circular fashion production.

1.2 Concept of Circular Fashion

The core concept of circular fashion involves full utilization of all the materials and products available in the society and their circulation among maximum users as well as for the longest possible duration. The materials used to create circular fashion are environmentally safe, effectively used and distributed in justified way among its users. There is no possibility of waste generation or there is no existence of waste in circular fashion, rather waste from one process is utilized as resource or nutrient for other production and consumption processes. Natural resources, virgin materials and renewable resources are used very efficiently and the use of new and virgin natural resources and materials is kept at smallest amount to curtail their undue exploitation and priority is given to renewable and reusable resources like recycled water, renewable energy sources and regenerated raw material to minimize the adverse impact of production and consumption on the environment. In addition to this, all the material used is free of all kinds of toxic chemicals and hazardous substances, in order to increase the circulation of safe and pure material among its users. Moreover the particles having risk of bioaccumulation at any level of the ecosystem are also not acceptable in circular fashion production to keep away the society from harm.

Perception of circular fashion essentially differentiates the biodegradable components and nutrients with synthetically manufactured components. Natural or biodegradable components obviously decompose in nature and work as compost, manure or nutrient for the soil whereas the synthetic components are non-compostable and non-degradable. Two types of cycles work in circular fashion—Biological cycle and Technical cycle. The natural biodegradable materials viz., cotton, silk, wool, viscose and wood fibers are considered as biological nutrients. These are entirely different and should be separated from the technical components which are non-degradable, such as polyester, nylon, acrylic, metals and plastics. The technical components can be recycled and streamed further in other processes of fashion production.

Since the technical (synthetic) and biological (natural) components of any fashion product can be separated and used in different flows of production, the designing of products containing variety of material types must consider easy separation of their individual components to facilitate trouble-free separate cycles for reuse. This "design for disassembly" is one of the key principles of circular fashion that will allow replacement, repair, redesigning, upcycling, maintenance and recycling of individual components of fashion products at the end of their use. The notion of circular fashion is simply based on the principle of disassembly of design and reuse of product. There is no existence of waste as well-thought designs are created considering reuse of waste from used fashion products at the end of their one life. These product cycles should be very stringent in their waste utilization processes to minimize waste generation and huge amount of raw material, energy, water, labour and other resources can be saved from being waste. Minimum utilization of virgin resources through reuse, maintenance, refurbishment and remanufacturing not only reduces production time and capital investment but also minimize generation of greenhouse gases, toxic chemicals and other hazardous byproducts. It implies the designing, production and manufacturing of fashion products for high durability, longevity and functionality besides ease in care, maintenance and repair with the purpose of maximizing its use and reuse in the ecosystem through technical and biological cycles.

In circular economy manufacturers and retailers are not only goods and product providers rather they work as service providers by following functional service model, so that the consumers not only use the products but also retain the products as long as possible with them by utilizing a range of services such as sharing, swapping, repairing, mending, refurbishing, reuse, etc. This kind of business model has shifted early disposable fashion products towards long-lasting ones. These models bring the customer towards more durable products with multiple uses to add to the functional lifespan of the fashion product.

With an intention of maximizing product longevity and durability, different design techniques and resources are to be sought out. Different services such as redesigning, recycling, repair are also required to be propping up by various fashion companies. The fashion corporations affianced with circular fashion system are expected to offer various repair and maintenance services for the customers—it can be in form of providing service or repair kits to the customers for home use or the required mending or repair can be done in the company stores as well. Redesigning is one more successful venture growing in India where the discarded fashion products are converted into inventive and transformed fashion garments. These products are custom-made and are closely attached to the emotions and sentiments of the customers. Tailor-made redesigned garments; rental or lease services of fashion garments are some other aspects of circular fashion model. To support and enhance such dimensions in circular fashion innovative type of infrastructure, collaboration approaches and innovative business models and techniques are essential to establish. Therefore, the circular fashion confides on cross-sectoral collaborations and partnerships between different businesses to facilitate efficient logistics and provide different services such as leasing, swapping, second hand, sharing, repairing, mending, redesigning, remanufacturing and recycling to extend the life of fashion products to give support to longevity and durability. Provisions of such collaborative services are the key feature of circular fashion for redesign, reuse and uphold circular flow of the fashion garments among many users.

Innovative design practices and production processes with effective and efficient use of biodegradable raw materials, non-toxic substances, renewable energy, water and other resources in the closed loops along with textile recycling are different features essential for the rise of circular fashion industry in India. Contrary to the linear "take, make and dispose" model of consumption where huge amounts of natural and virgin resources are exhausted and shattered unintentionally, the circular fashion model work towards efficient utilization and consumption, that is, minimal consumption of resources so as to preserve them for future generations. Indian fashion industry needs to change its predominant business model of "take, make and dispose" towards the circular model where long-lasting products are designed and manufactured in small and demanded volumes for active and favourable use for long periods of time by its one or multiple users; the components being able to be recycled many times for multiple users, lowering the cost of production and maximizing the profit of the product manufacturers. Therefore, the manufacturers and retailers should maintain the possession of their products for as long as possible by providing different service and multiple ways of consumption. They should develop efficient and effective collect, recollect and recant system and produce more durable products that can be disassembled and refurbished easily and appropriately to increase their active number of life cycles.

Hence, the umbrella of circular fashion has three major canopies—Design, Production and Use. The fashion product, whether it be an apparel, an accessory, home furnishing or any other particle, can be defined as a circular fashion product only if it confirms to the following parameters:

1. Design—The item should be designed in such a way that its component parts can be disassembled or separated for the purpose of mending, repairing, reconstruction and reuse during its use as well as recycling at the end of use as indicated by the requirement. The designing requires incorporation of high-quality, non-toxic raw material with enduring, classic and everlasting designs and styles to take full advantage of its durability, longevity, exquisiteness and attractiveness besides its utilization by many users to make best use of the resources used to produce the item. The high-quality items are regarded as prestigious belongings as they look good and best utilized by the consumers. The design is based on specific requirement of the customers, i.e., customized design because such designs are closer to the customers' emotions and maximum utilized by the consumers and they want to use it even after redesigning, thereby increasing its lifespan and worth. Such authentic and genuine items are intended to use actively for their whole lifetime by the consumers.
2. Production—The fashion item should be produced with high-quality, non-toxic and biodegradable materials so that its constituent materials can be composted or biologically degraded at the end of use or if it is produced with non-degradable or synthetic materials, it must be suitable for recycling efficiently. Consumers are, nowadays, concerned with various health and environmental issues so they look for associated certification such as Eco labels and GOTS, etc. The production process requires minimum waste generation preferably utilizing least amount of virgin resources, allowing the leftover part to be reclaimed and reused as raw material and nutrient for other processes. Production, transportation, distribution and marketing of the product incorporating very efficient and safe use of raw materials, renewable energy resources, recycled water and other resources is ought to be preferred wherever possible. The consumers prefer to purchase items that are made with environmentally certified organic and natural materials such as cotton, linen, silk, wool, etc., over synthetic and toxic material like polyester, nylon or acrylics. Similarly the recycled materials, ethical and fair trade products are more favoured by many vigilant consumers in today's world.
3. Use—The quality and attributes of the fashion product be supposed make it useful for multiple users thereby extending its practical life through sharing,

renting, swapping, borrowing, mending, repairing, redesigning and second hand services. At the end of one life cycle, the components of the product ought to be recycled and reclaimed safely and effectively, and after use the components can be utilized for remanufacturing of new products or through biological decomposition they turn into microorganisms and provide nutrition to the soil. The circular fashion item is in opposition to the linear one, where each and everything is utilized, reutilized, purposed, repurposed, reclaimed and recycled in the best possible, most efficient, sustainable and ethical way and nothing goes to the waste and landfill to increase the environmental pollution.

Circular fashion is an element of sustainable fashion. The concept of circular fashion is based on closed-loop circular economy to restore and regenerate existing fashion products. The sixteen key principles of circular fashion concern the entire life cycle of a product, from design and sourcing, to production, transportation, storage, marketing and sale, as well as the user phase and the product's end of life. The circular system, based on circular economic principles, is a way to generate more income, to create more profitable and sustainable businesses as well as minimizing the negative impact on the environment of our planet Earth. Circular fashion can counter to the prevailing linear model of "take, make and dispose" in the textile and fashion industry.

Circular fashion can be defined as clothes, shoes or accessories that are designed, sourced, produced and provided with the intention to be used and circulated responsibly and effectively in society for as long as possible in their most valuable form, and thereafter return safely to the biosphere when no longer of human use [8]. In a broader way, it can be stated that circular fashion products are designed with ethics, possess high longevity, and are resource-efficient, non-toxic, biodegradable, recyclable and reusable. Likewise they are produced with locally available raw material in safe and ethical environment. In addition the circular fashion products are used for multiple life cycles, for a long time in good condition through proper care and maintenance, repair, refurbishment and sharing among multiple users. In the circular fashion model, the products are redesigned to reuse and recycle the material and components to manufacture newer fashion products. If the constituent material is not suitable for recycling or reuse, it is composted in the soil to supply nutrients. On the whole, the circular fashion products do not have any negative environmental or social economic impact (Fig. 1).

Circular fashion contributes to positive development and well-being of humans, ecosystems and society at large. It increases job opportunities as it emphasizes more on human labour rather than energy consumption for fashion production which in turn reduces the exhaustion of natural virgin resources and negative environmental impact either having multiple life cycles or biodegradability. Besides it facilitates economic competitiveness between producers and appearance struggle among consumers by designing the products according to their personal preferences. A move from linear fashion to circular fashion involves increasing the value and lifespan of fashion products, components and materials multiple times, elimination of waste throughout the supply chain and thereby reduction in environmental impact of the apparel industry

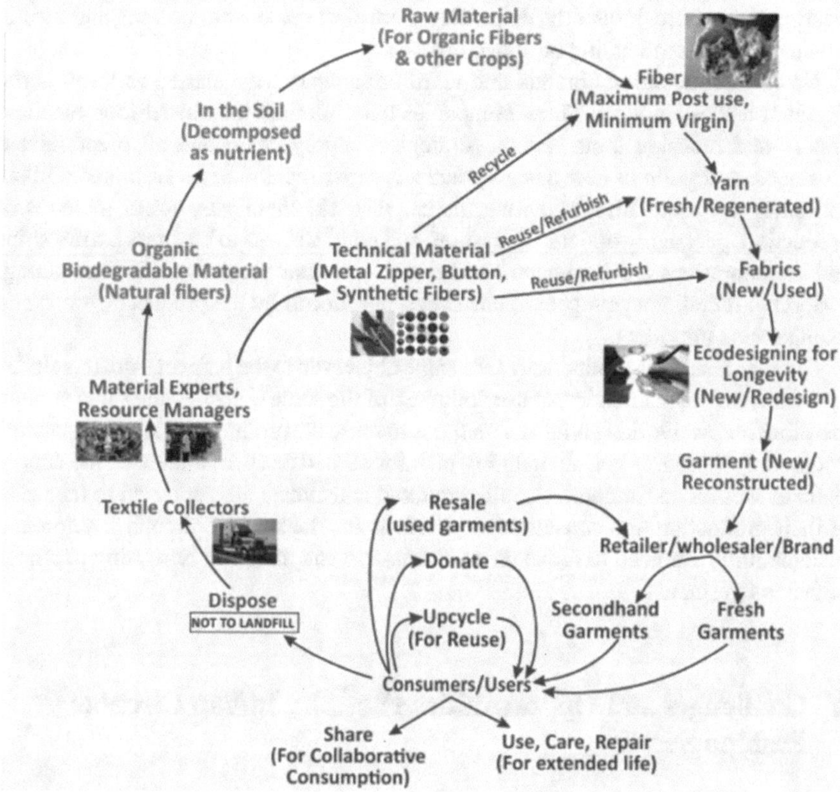

Fig. 1 Circular fashion in textile and apparel industry (*Source* The Author)

as well as restoration, regeneration, re-establishment and protection of natural virgin resources.

The cradle-to-cradle approach of circular fashion model permits fashion producers to use the textile waste to create new fibres, yarns and accessories from non-degradable material and soil compost from the residue of natural textile fibres [16]. This is aimed at the repeated development and production of high quality and valuable products from discarded constituent waste of fashion products in a closed-loop system. Products are designed in a manner that all the textile waste from one stage of manufacturing process is used as nutrient or raw material for another stage of production and the resultant waste is zero. The flow of material occurs within the closed loop and recycling and reuse of the waste increases material recovery. As the resurgence involves both pre-consumer and post-consumer wastes, the circular fashion approach requires collaboration of all the stakeholders of fashion production and consumption, viz., designers, manufacturers, suppliers and even consumers. A significant feature of this paradigm shift is focused on consumers who can extend the life cycle of clothes during its use phase—by reducing purchase of the clothes, wearing them more frequently, discarding them after reasonable wearing and further reusing them after mending or upcycling.

Now in India various brands and retail companies have started to think about circular business models. These companies have adopted the closed-loop business system and redesign their visions, strategies, supply chain and all other related processes. Adoption of new design practices, experimentation on sustainable fibres and other raw material, eliminating the use of toxic chemicals, newer methods of production, packaging and transportation, and new business to business partnerships and collaborations are instigated. Additional customer services such as mending, swapping, rental, upcycling, etc., have been introduced by the producer groups and retail service providers.

Circular fashion is production of the same objects up to the point of sale. It includes extension and intensification of use and reuse of the same object, in the same context therefore innovative designing and methods to reuse the products are key components of circular fashion. It is a challenging task for all textile stakeholders of the supply chain as well as consumers. Manufacturer and consumers are supposed to take care of their production and consumption practices for the sake of circular fashion and sustainability. We need to focus on conscious use and adoption of reusing practices of fashion garments.

2 Challenges and Opportunities Faced by Indian Circular Fashion Sector

The Indian circular fashion sector is facing many challenges as they persist worldwide. India is not only a big consumer of textile products but also it is a manufacturing hub for fashion and textile products. The technological advancement and improving living standards of Indian populace have given rise to awareness and concern for environment and sustainability. This, in turn, creates various challenges in front of the producers and consequently open new vistas and opportunities for them to innovate and to attain new horizons of their businesses.

The supply chain of fashion industry accounts for 4% of total global greenhouse gases emissions. This negative impact of global supply chain is a result of massive use of raw material, inconsiderate water consumption, thoughtless energy emissions, injudicious waste creation and heedless chemical use. To address these challenges, the industry needs to adopt innovative way-outs and radical changes in the production process as well as throughout the supply chain.

The incremental, well-efficient sustainable and target-focused innovative initiatives are required to be set by the players in the fashion industry and the attainment of progress requires shift from the contemporary consumption and production patterns towards restorative and regenerative utilization and manufacturing process to reach the innovative goals of circular sustainable fashion. For this, the industries have to set the objectives to integrate sustainability and innovative circular fashion techniques in their core business operations to meet ecological targets of the society in conjunction with averting any financial adversities or economic losses.

The increasing awareness among consumers and stringent governmental policies regarding social and environmental concerns associated with fashion production and consumption have accelerated the sustainable and circular drive of fashion industry. Strict ecological regulations and conventions in production policies to incorporate circular movement have stipulated the industry to a great extent to go forward for circular production, bring changes towards innovative solutions, improve resource efficiency, radical and divergent thinking as regards to production processes and business models during the last decade. These models and policies have significant impact in decreasing carbon footprint of fashion production, distribution and consumption as well.

Dr. Anna Brismar [8] has identified sixteen key principles to support and promote a more circular and sustainable fashion. The principles basically relevant for producers and business persons include design with a purpose; design for longevity, resource efficiency, biodegradability, recyclability; source and produce more locally, with efficiency, with renewable, with good ethics, without toxicity, to support long life, reuse and recycle and collaborate well and widely. Whereas the principles that are primarily concerned with the consumers are reuse and recycle; use, wash and repair with care; consider rent, loan, swap, second hand or redesign instead of buying new; buy quality as opposed to quantity. In order to adopt these principles in practice, the sector has to face many challenges throughout its supply chain.

Constantly changing consumer behaviour, their knowledge and awareness regarding social and environmental issues, national and international business policies, governmental outlook over the prevailing business practices are compelling the producers towards circular fashion production. Strict environmental regulations and policies are forwarding the industry towards sustainable solutions and innovative practices. It includes resource efficiency, sustainable processing and manufacturing techniques as well as refurbishing and recycling of the fashion products. This has significantly reduced the adverse effect of apparel sector ecosystem and biodiversity, minimized the footprint of textile production and distribution in addition to the consumers' habits of clothing purchase, use and maintenance.

To meet the challenges emerging constantly, many players of the industry have shifted their supply chains radically with efficient focused incremental, restorative and regenerative innovations to attain their targets and integrated sustainable solutions in their core business operations that will lead to financial and environmental benefits to the global society. New business models are emerging to facilitate supply–demand side competitive advantage to the industry through circular economy. These models include various ways, such as circular inputs of raw material, recovery of useful resources from processes and discarded products, extending the useful life of products and services, opening up new ways of sharing the products and using the products as services for more than one consumer, and many more.

The challenges encountered and forthcoming opportunities in various stages of supply chain are discussed below:

2.1 Designing

Designing of the fashion products for their extended and multiple life cycles is a big challenge as this requires performance-driven designs rather than style-driven. Circular designing focuses on the performance of the product and its constituent material, the ways it can be reused with innovative designs. The cradle-to-cradle approach of circular fashion emphasizes on designing products with zero waste [19]. With this purpose products are designed with keeping their end in mind. The design should not only be good for people and environment but also the material used needs to be reusable; the waste generated from one product should essentially be the part of another one. The challenge is to create designs that can be recycled or refurbished after their primary use and continuously circulate within the closed loop of circular fashion. Zero waste is essential to put away from design failure. Material reutilization is key concern for the designers to proceed for circular product designing and the design should include only that material which can be either recycled within the industry or can be used as compost, but not even a smallest part of the generated waste goes to landfill and pollute the environment. The discarded waste is considered as the part of new product during designing. The designers need to go no in line with nature where there is no waste.

In India, our traditional clothing comprises of many unstitched garments that do not produce any pre-consumer waste and we have deeply rooted traditions of reuse, refurbishing, mending and upcycling leading to zero waste during production and use. Nowadays, many designers and companies have adopted circular fashion principles. They design formal and informal outfits in such a way that the consumers cannot readily differentiate the fresh and upcycled garments. They work in a regenerative system where the garments are designed and redesigned to be circulated for longevity to maintain their ultimate value during their lifespan and thereafter when they are of no use to be recycled or decomposed safely in the biosphere.

Circular design indicates that every part of any fashion product can be used in a cyclic way—each and every part of the garment is timeless, durable and can be used multiple times. The designing is supposed to facilitate disassembling of component pieces and other parts of the garment at the end of one useful life, so that the parts can be reused for the purpose of reassembling, remanufacture and reuse to initiate a new verve. The circular model of designing considers ethical and fair creation of the fashion product. Products are designed and developed with the concept of resource efficiency, durability, non-toxicity, biodegradability and recycling; priority is given to the recyclable and ethical resources to produce fashion garments and sustainable production processes keeping their next use in mind. When the garment reaches the end of its one life, circular designing makes its repairing or redesigning possible rather being disposed and go to landfill. It can be shared, rented, swapped or sold at second hand shops. This endow with less exploitation of natural virgin resources available on the planet. If any part of the product is inappropriate for recycling, it should be biodegradable to be used as compost for the soil to provide nutrition for other plants and organisms of the ecosystem. Hence, the circular design system gives a way for fashion to be self-sustained and self-reliant as an alternative to exploit and diminish the natural resources available on the earth.

2.2 *Raw Material*

After designing, the next step is to collect raw material for the production of fibres, yarns and fabric. More than half of the environmental footprint of textile industry is generated in this stage. Therefore, selection and collection of raw material requires alternative and improved raw material to create fashion products. The challenge is to discover alternative plant sources of fibres and assess them for spinnability and other desired properties to transform into apparel. Even in recycled polyester or other synthetic fibres, it is very important to be well aware of its source and the chemicals used in the process of recycling. More innovation and capacity building is required to increase reuse of many natural cellulosic and synthetic fibres.

In India, traditionally, low-impact natural fibres are used commonly for yarn and fabric manufacture. Innovations and experiments are under process for massive use of low-impact fibres. These can be extensively used alone or with manmade cellulosic fibres at industrial level for mass production [1]. Several Indian fashion players have incorporated recycled fibres, fabrics and garments into their collections for commercial purpose. Plant-based innovative natural fibres, banana fibre, lotus stem fibre, fibres from leaves, fruit skins and agriculture and food waste products like straw, wheat husk and coconut husk are being explored for apparel use. Regenerative protein fibres and biosynthetic fibres are nascent and relatively unexplored in India for fabric production [22].

Investment to improve processing technology of alternative fibres, creation of formal supply of agro-waste, public and private partnership to develop local supply chain and maximum involvement of locally available human resources for fibre and Fabric production can give impetus to circular fashion production.

2.3 *Manufacturing Process*

The wet and dry processes for fibre, yarn and Fabric production, dyeing and finishing require lots of toxic chemicals, water and energy. Moreover, a huge amount of hazardous gases, poisonous and noxious wastes are released in the environment that significantly pollutes surrounding air, water and soil [3]. The chemicals widely used in wet processing of fabrics are not only perilous to the environment but also dangerous for human health. The challenge is to adopt more sustainable pretreatment, dyeing and finishing processes, to produce fabrics without use of any toxic chemical along with minimum water and energy consumption.

Many producers have adopted sustainable processes and innovative techniques in natural dyeing and printing commercially, though they are not on a larger scale [24]. Some of the Green Chemistry solutions prioritized by the players are—use of plasma in pretreatment and finishing processes, ultrasonic and foam dyeing, nano-dyeing, use of supercritical carbon dioxide in dyeing, microbial pigment printing, spray printing, cationic and enzymatic finishes. These solutions have reduced water and energy consumption in wet processing of materials and decrease in wastewater pollution as well. Besides digital printing, 3D printing has replaced to some extent to the water and energy exhausting processes [10].

The projects concerning renewable energy, water stewardship and carbon management have been started by some manufacturers that will lead to minimum consumption of these resources and reduce water and environmental pollution. New sustainable environment friendly products and processes for textile manufacturing are to be sought out to bridge out the gap between research findings and their implications at industrial level [27].

Formally organized dyeing and printing sector that can be motivated towards adoption of sustainable and non-toxic processes, sludge treatment and water purification with the help of government agencies, industry academia linkages for textile processing can assist in meeting about the challenges occurring in this stage of production.

2.4 Garment Construction

Today shaped garments are more prevalent in India instead of draped clothing. Cutting, stitching and trimming of these apparels produce a huge amount of waste [9]. The challenge is to use and reuse this waste in the closed loop. Acceleration of Mass customization digital designing and e-commerce platforms are challenging tasks in the Indian apparel sector. Additive manufacturing is gradually creating its way in Indian apparel industry though it was existed earlier in low-tech forms of hand and machine knitted product and crocheting that needs improvisation with advancing technology [12].

Innovations in cutting, stitching and trimming are in budding stage in the Indian apparel sector. Innovative technologies to eliminate any kind of waste during garment construction, for example, zero- waste designing and minimal seam constructions are being introduced in this sector. Zero-waste pattern cutting and stitching are in nascent state in India but these can amazingly minimize flooding of textile waste into landfills and incineration. Slow fashion is an emerging concept among Indian consumers which provides high-quality clothing with longevity and ethics.

Optimization of construction can increase the overall functional performance of yarn and Fabrics. Garment workers can be upskilled to adopt innovative techniques in cutting, stitching and trimming for apparel manufacturing; funding for 3D printing and zero-waste cutting at large scale can provide solution to waste generation [5];

development of tools and equipments are big opportunities to accelerate circular fashion.

2.5 Packaging and Transportation

Single-use packaging material made of plastic, polythene, paper or cardboard are extensively used in textile and apparel packaging that considerably contributes to high carbon emissions in the environment. Plastic and poly bags used for packaging of the garments add to non-degradable waste generation. Innovative packaging solutions such as eco-friendly packaging material, recycled materials, reusable packaging, biodegradable material and bioplastics are though in experimental phase, create an innovative opportunity for the packaging of textile and apparel products in India and round the globe.

Both the global and domestic supply chains require frequent transport of raw and finished products. Our apparel sector has to meet out logistical challenges, especially in the e-commerce segment where operation of business activities take place worldwide. Use of locally produced raw material and selling the products in local markets to minimize transportation and circular logistics are issues of concern for manufacturers and suppliers.

2.6 Retailing

The textile sector is producing more than global clothing requirement; therefore a large quantity of textile products remains unused and further generates solid waste. Many unwanted apparels are discarded as trash or sold as second hand clothing. But again, only a small portion of second's clothing is used and the remaining adds to increase the volume of landfill.

Retailing is that one stage where environmental impact of clothing industry can be reduced to a great extent by encouraging slow fashion, customized tailoring and retailing, facilitating mending services, clothing rental or swapping and upcycling services. In the Indian apparel sector the pace of new garment production can be reduced extensively resulting into trimming down emission of carbon and hazardous gases as well as curtailing virgin natural resources required in textile manufacturing process. The challenge is to change the mindset of customers regarding multiple uses of fashion products in addition to improvise services concerned with redesigning, upcycling, rental or second hand clothing.

Extension of useful life and increase in number of life cycles of textile products have wide prospect in India. Though the market for rental garments, second hand clothing and swapping activities is limited to luxury and formal garments, it is expanding swiftly towards casual wear too. At this platform dead stock of luxury garments is primarily being utilized for or multiple wearing. Innovative services are being provided to create an ecosystem for mending services, redesigning and upcycling that gives extended life to the garments. Technological advancements like virtual reality, digital solutions and e-commerce platforms are providing the retailers with efficient marketing strategy and successful approach to the customers to satisfy their personal requirements. Various circular business models pertaining to re commerce, rework and rental are being adopted in India. The rental model is gaining remarkable success in luxury and formal wear segments. The rental market is thriving speedily in India and it is expected to grow 22% by 2027 [13].

Players in textile industry have created ecosystems to convert pre- and post-consumer textiles into new accessories plus household and other decorative articles to increase the life of textiles. It is not only beneficial for environment but also provides employment to the local people. Rework, repair and customized tailoring have traditionally been a part of Indian lifestyle. These were greatly replaced by ready-to-wear garment segment during 1990s but they are regaining popularity nowadays [7].

Formal organizations of the service-based enterprises and customized tailoring shops can perk up these sectors and hence assist to augment their in-time services, delivery at home, pick-up and drop-off services to boost up the market of circular fashion. The garments having short-time use like maternity and children's wear have significant opportunities in rental and e-commerce markets. Companies can provide customized services and incentives for their specific customer profiles to encourage them to extend the life of their clothes.

2.7 Consumption and Use

Lacy [18] believed that circular economy is no longer just about the supply side, but is becoming much more of a customer proposition. Rapidly changing fashion, fast consumption, early disposal, never-ending search for new clothing design, impulsive buying are major consumer practices that consistently produce huge amount of landfill damaging the environment and Planet. The challenge is to educate the consumer regarding negative impact of fast fashion as well as creating awareness for sustainable and cautious purchase, proper care and maintenance for extended life of the apparels. Besides, stigma attached to second hand, rental or upcycled clothing needs to be removed from the psyche of consumers.

Generation Z is well aware of e platforms for rental, swap or leased garments. The disgrace and shame associated with used or second hand/rental clothing is gradually fainting showing multiple ways of closing the loop. Sharing of garments has been a part of our lifestyle from time immemorial. These factors favour expansion of circular fashion products, and they are marketed in India. Indian consumers today possess consciousness and intelligence regarding purchase and use of clothing. They are adopting new practices to use the existing clothes by more wearing and maintaining them well. Investment focusing towards smaller and smarter wardrobe by means of less clothing with varied and multiple combinations are the new trends. This is called a curated wardrobe based on wiser purchase decisions whereby each garment is seen as an investment and thus wardrobe content is constructed slowly [14]. This kind of clothing choices can substitute and provide an alternative approach to fast-fashion consumption.

It is a mistaken belief that quality of recycled and upcycled goods is inferior in comparison of the goods made up of virgin material. There is wide scope to produce better quality yarns and fabrics from recycled or upcycled fibers.

2.8 *Post-consumer Use*

Any fashion product that is disposed goes to the landfills or for incineration as they are not designed for multiple life cycles or circularity. A big amount of their constituent material can either be reused, recovered or recycled. Approximately 25% of the used garments are collected globally for reuse and recycling and only 1% of the recycled garments are converted into new materials [2].

Recycling of post-consumer waste is a difficult task as sorting, grading and transportation, logistics are issues of major concern at this stage. Sorting of the disposed material according to fibre type and segregation of embellishments and fasteners are big challenges faced during recycling process. Smaller recycling units have to face more problems because recycling of blended fibre fabrics require hundreds of kilos per fabric per colour to run one cycle of shredding, carding and spinning. Automated sorting technology is yet to be explored and adopted widely.

As the post-consumer waste for recycling is imported from Europe and the United States, and the finished product are exported outside India, efficient collection systems are utmost required. Whereas the pre-consumer waste is collected from domestic market and down-cycled as low-cost material like mops, blankets, bedding, etc., and sold in domestic market. The challenge is to collect and use local post-consumer textile waste for recycling and creating a market within the country for recycled products.

The products made by chemical recycling are compatible to virgin products in quality and have very low impact on environment, therefore chemical recycling should be promoted to create fibre, yarn and other textile material. Though the complexity of technology and high capital investment are among few impediments in chemical recycling, it has immense potential in Indian textile sector. Collection, repairing and reselling of second hand clothing from households at local level is being done for a long in India that lend a hand in extending product life by making low-cost garments and other textile products. Many domestic and global brands have shown their interest in recycling of textile products and converting them into yarns and fabrics.

Recycling PET bottles into polyester fibre is the only major area in the field of recycling. Inventive techniques of chemical recycling for domestic post-consumer waste, automated sorting technology to intensify speed and scale of recycling, creating an ecosystem for recycled products have prospects in utilizing post-consumer domestic textile waste. In India, mechanical recycling is more popular and gaining impetus as compared to chemical recycling of post-consumer textile waste. India is global hub for mechanical recycling with the processing of 5 million tons of material [25]. Strong collaboration among various suppliers and recyclers is required for maximum recycling of the pre- and post-consumer waste, especially at the domestic level, as the recycling segment is in its growing stage and not fully and formally organized yet.

3 Conclusion

The linear system of fashion consumption is detrimental for both mankind and the environment. Polluted environment, climate change, global warming, depletion of natural resources, escalating skin and other diseases are some of the drawbacks of fast fashion faced by the society.

Indian textile and apparel sector is showing a transit towards circularity. Integration of circular business models or any aspect of circularity in any stage of garment production from designing to recycling can assist closing of loop. Despite many challenges, opportunities for circularity are apparently perceptible as rising consciousness of Indian consumers is propelling the industry towards sustainability. Innovations in circular fashion sector have potential to enable the industry to practice efficient measures for circularity. Though the degree of momentum varies significantly at different stages of supply chain, adoption of circular practices by all the stakeholders of textile value chain including end-consumers because of uniqueness of Indian landscape, the opportunity of circular fashion in India are enormous.

Consumers can play a key role for transformation from linear to circular fashion that leads to regeneration of the natural resources, curtail textile waste and carbon footprint of the apparel industry. Redesigning of supply chain, empowerment of workforce, upskilling of labour, technical assistance to the workers, innovative and efficient techniques of logistics and packaging, digital acceleration from designing can boost up closing of the loop. It calls for bringing together industry practitioners, academic institutions, research and innovation centers, civil society and governmental agencies on one platform to meet out the challenges and implementing circular solutions to exploit completely the opportunities and future prospects associated with circular fashion.

Thus circular fashion furnishes textile and apparel industry with innovative and sustainable paradigm to contribute to positive development and well-being of humans, ecosystem and society at large. Circular fashion can transform the biggest pollutant fashion industry into the most ethical and sustainable one with zero waste.

References

1. A green opportunity for viscose processing in India. http://www.sustainabilityoutlook.in/content
2. A new textiles economy: redesigning fashion's future. https://www.ellenmacarthurfoundation.org
3. A study on the effects of pre-treatment in dyeing properties of cotton fabric and impact on the environment. https://www.hilarispublisher.com/open-access
4. Armstrong C, Niinimäki K, Lang C (2016) Towards design recipes to curb the clothing carbohydrate binge. Des J 19(1):159–181
5. Automation to affect employment rate in Indian textile industry. https://apparelresources.com
6. Baker-Brown D (2017) Resource matters. In: Baker-Brown D (ed) The re-use atlas. Riba, London, pp 7–15
7. Be spoke tailoring marks a comeback in the Indian Fashion Entity. https://www.indianretailer.com
8. Brismar A (2017) Unfold your circular potentials: supporting a more circular fashion industry. https://circularfashion.com
9. Creating fashion without the creation of fabric waste. https://www.researchgate.net
10. Digital printing—Shree Ganesh Flex Printing. http://www.shreeganeshflexprinting.co.in
11. Ellen MacArthur Foundation (2017) A new textiles economy: redesigning fashion's future. https://www.ellenmacarthurfoundation.org/assets
12. Evolving IT trends for mass customization in apparel industry. https://www.fibre2fashion.com
13. Fashion rental catching up pace. https://www.indianretailer.com
14. Filippa K (2016) The curated wardrobe: the recycled cashmere knit. http://filippakcircle.com/2016/12/14
15. Fletcher K (2008) Sustainable fashion and textiles: design journeys. Earthscan, London.
16. Herranz AG (2017) Closing the clothing loop: a cradle to cradle platform for fashion. https://en.reset.org/12.12.2017
17. Kant R (2012) Textile dyeing industry: an environmental hazard. Nat Sci 4(1):23
18. Lacy P, Rutqvist J (2015) Waste to wealth: the circular economy advantage. Palgrave Macmillan, Hampshire
19. Lynggaard, H (2017) How to design for a closed loop: Cradle to Cradle® in Fashion June 29, 2017. https://en.reset.org/
20. McAfee A, Dessain V, Sjoeman A (2004) Zara: IT for fast fashion. Harvard Business School Publishing, Boston
21. Niinimäki K (2011) From disposable to sustainable: the complex interplay between design and consumption of textiles and clothing. Doctoral dissertation, Aalto University, Helsinki
22. Regenerated protein fibres: a preliminary review. https://www.researchgate.net
23. Runnel A, Raihan K, Castel N, Oja D, Bhuiya H (2017) Creating digitally enhanced circular economy. Reverse Resources. http://www.reverseresources.net/about/white-paper
24. Sustainable approaches in the dyeing and textile industry in India. https://communities.acs.org
25. Textiles recycling: rising pressure on India's recyclers. https://www.recycling-magazine.com/2017/10/20
26. Valuing our clothes: the true cost of how we design, use and dispose of clothing in UK. http://www.wrap.org.uk
27. What's coming for the dye and dyestuffs industry in India. https://www.meghmaniglobal.com

A Study on the Comparison of Fabric Properties of Recycled and Virgin Polyester Denim

J. Swetha Jayalakshmi and D. Vijayalakshmi

Abstract Each year, nearly 8 million tons of plastic waste enter the oceans and it was predicted that this causes a serious climate change. Even though fashion is ephemeral, producing products less toxic helps in protecting the environment. Being a large-scale sector, it is the responsibility of the apparel industry to be sustainable. Denim creates a fashion statement. People of all age groups like denim due to its comfort and looks. Even though it is one of the oldest fabrics, it remains an evergreen classic. Denim is used in many areas, not only trousers, but it is being used for making jackets, shirts, blouses, accessories, etc.

In this study, an attempt was made to compare the fabric parameters of polyester denim with recycled polyester denim. Tests such as GSM, tensile strength, shrinkage and colorfastness were analyzed. The strength and shrinkage results were similar whereas the colorfastness results were the same for both the fabrics. Thus, this study demonstrates in using recycled polyester as an alternative for virgin polyester as there are no drastic changes between the properties of both the fabrics.

Keywords Denim · Environment · Recycled polyester · Shrinkage · Strength

1 Literature Review

1.1 Denim

Denim, the favorite fabric of the youngsters has indeed come a long way. The word denim is an Americanization of the French name "serge de Nimes," a fabric which originated in Nimes, France during the middle ages [1]. Denim is evergreen fabric in the fashion cycle. They have diverse advantages compared with other

J. Swetha Jayalakshmi (✉) · D. Vijayalakshmi
Department of Apparel and Fashion Design, PSG College of Technology, Coimbatore, India
e-mail: jsj.afd@psgtech.ac.in

D. Vijayalakshmi
e-mail: hod.afd@psgtech.ac.in

S. S. Muthu (ed.), *Sustainable Approaches in Textiles and Fashion*, Sustainable Textiles: Production, Processing, Manufacturing & Chemistry,
https://doi.org/10.1007/978-981-19-0530-8_4

types of clothing. They are comfortable, adventurous, relaxing, attractive, aggressive, smart, casual, dynamic, energetic, aesthetically appealing, timelessly fashionable or creative. This explains why generation after generation finds the denims as a material of first choice for casual wear. Modern consumers are interested in clothing that not only looks good but also feels great [2].

1.2 History of Denim

The denim fabric was used for jeans and first for the miners, ranch hands and farmers. The "twill weave" which is accepted because it made dirt and stains less visible, flexible and comfortable. It is produced by yarn-dyed spun yarns of cotton or its synthetic blends. It is a characteristic of most indigo denim that only the warp threads are dyed, whereas the color of weft threads remains plain white. Due to warp-faced twill weaving, one side of the fabric shows the colored warp threads whereas the other side shows the white weft threads [3].

The traditional denim is usually hard and high density fabrics with high mass per unit area. Mostly twill weaves such as three-up-one-down (3/1) and two-up-one-down (2/1) are used for denim construction. Denim is available in attractive indigo blue shades and is made for a variety of applications and in a wide range of qualities. Denim's durability lies in the combination of the yarn and the weave. The consumer's today need durability and comfort in their fashion items including denim. Twill weave fabrics absorb a lot of friction before it breaks apart thus having a good abrasion resistance. The way in which the yarns are woven together is the reason for excellent durability.

1.3 Applications of Denim

Denim is a good choice for casual jackets, skirts and jeans. In clothing denim is used as capri pants, jeggings, suits, dresses, shorts, sneakers, dungarees, shirts, kurthis, etc. Other than clothing, denim is also used in making accessories like tote bags, handbags, footwear, belts, etc. In furniture, denim is used to make bean bag chairs, lampshades and is also used as upholstery material.

1.4 Different Washes and Types of Denim

Denim is normally dyed with indigo, vat and sulfur dyes. Among these, indigo share is 67% [4]. One of the widely used finishing treatment that has vast usage in textile sectors is denim garment washing. Without finishing treatments, denim garment is uncomfortable to wear, due to its weaving and dyeing effects. Finishing helps in making the fabric softer, suppler, smooth and comfortable to wear performance. Popularity of denim washing in the world market has been increasing day by day. There are introduction of new finishing treatments of denim garments to meet up the trends of the fashion industry. The most common denim washing methods are bleach wash, acid wash, enzyme wash [5], normal wash, stone wash, etc. Bleach method is widely used method in industry especially for denim washing to achieve required color shade by hypochlorite bleaching. It is observed that in wet processing chemical usage is more [6].

Various levels of color removal with little or no damage to the denim material are achieved by using different laser parameters. There are various advantages of using laser-based finishing treatments such as less pollution of the environment by reduction of chemical agents and water consumption [7]; process flexibility allows replication of existing stonewash designs or creation of new styles; automated control of laser-faded technology [8].

Using an ozone wash for denim instead of a hypochlorite or PP bleaching recipe can help in reducing water usage, as fewer rinse cycles are required, which thereby reduces the energy required to heat rinse water. It can also eliminate the use of dispersing chemicals to remove back staining in pockets and weft yarns. It can also reduce batch-processing time by eliminating rinse and neutralization steps required for bleach or PP, and also reduce damage to and loss of denim fiber strength.

Ozone bleaching can provide many advantages when used alone or in combination with traditional chemical bleaching methods for denim finishing. Water and energy reductions in the range of 50% have been reported. It can also reduce the total amount of chemicals used by eliminating bleach and PP spray from the process. When ozone is combined with laser technologies, a more sustainable product is created. By combining these innovations, we can achieve major reductions in our use of water and harmful chemicals. This is not only a benefit for the environment; combining these resources also creates a safer environment for factory workers.

1.5 Sustainability in Apparel Industry

In the current economic system, the fashion industry has an important part. Here is where the skilled handicraft is combined with technology to develop end products [9]. The apparel industry is one of the intrinsic parts of the Indian textile sector [10]. Being a large-scale sector, employing millions of people, these industries remain the second-largest polluter in the world. In each step of garment production, there are several steps in making it sustainable. Now, companies are striving for an eco-friendly way to meet the severe environmental crisis. Consumers are also interested in sustainable living. Adapting constant changes is the necessity to survive in any market. Thus companies incorporate sustainable strategies into their business practices.

Denim is a versatile garment. From kids to adults everyone likes this fabric due to its comfort and looks. Usually, denim needs a finishing treatment to achieve comfort. Indigo dyes are being used from olden days. The warp yarns are dyed while the weft ones are undyed. Blended denim fabric provides designers to bring innovation in fabric development [11]. It is seen that a classic levis 501 jeans take 33.4 kg of CO_2, 3.781 L of water and 400.1 mJ of energy for the production which is equivalent to 246 h of watching TV on plasma big screen, driving 69 miles. Denim is considered as the topmost water-consuming textiles during production [12]. Nearly 15% of denim waste is collected during the cutting process [13]. Similarly, wastages happen in all stages of garment production. Thus each stage in apparel industry is considered to find ways to impart sustainable methods.

The demand for textile products is always on an increasing scale. Some studies reveal that nearly 63% of fibers are based on petrochemicals and 37% is based on the thirsty plant "cotton." There are several studies done on recycling and reusing materials, from a review paper it was seen that nearly 85% deal with recycling and fiber recycling was done for 57% which contributes to a major scale. It was also noticed that cotton and polyester are mostly studied fibers which contribute to 76 and 63%, respectively [14]. Hence, these studies claim to do recycling which in general less harmful when compared to landfills, incineration, etc.

Polyester is a fiber that depends on utilizing non-renewable resources [15]. The process of recycling PET waste is a well-established system [16]. Studies reveal that there is a lack of using recycled polyester in the products of fashion brands [17]. By performing sustainable processes, there come several benefits such as it reduces the cost of purchasing goods, minimizes the wastages and reduces environmental hazards, increases profit and decreases the cost of disposal and treatments [18]. Recycling is one of the eco-friendly processes. Recycled fibers can be blended with other fibers to enhance their properties [15]. The usage of elastane helps in freedom of movement and ease in denim [19].

1.6 Sewing Parameters

A seam is considered as an essential measurable factor that makes up a garment. Seam strength and quality influence a lot in the service of the garment. According to ISO4916:1991, there are eight classes of seams namely,

1. Class 1—Superimposed seam
2. Class 2—Lapped seam
3. Class 3—Bound seams
4. Class 4—Flat seams
5. Class 5—Decorative/Ornamental stitching
6. Class 6—Edge finishing/neatening
7. Class 7—Attaching of separate items
8. Class 8—Single-ply construction

Class 2 of lapped seams are commonly used in jeans as it is very strong and it can take a lot of wear [20]. Based on the style of the garment, its functionality and appearance, there are different types of stitches used in garments. Among them,

Class-300 (Lock Stitch) is chosen [21]. In this type, the stitch formation involves a minimum of two threads locking together in the center of the plies of fabric being sewn. It is more secure than the 100 class as the tendency for the stitch to unravel is reduced due to the locking mechanism. Usually, the greater the seam strength when the stitches are greater. But in some cases, too many stitches can damage the fabric by cutting the yarns enough to weaken it [22].

When the sewing quality of the garments is considered, sewability is an important factor [23]. Generally speaking, the seam quality depends on the seam strength and appearance. Based on the end-use the seam strength may differ [24]. The type of sewing thread and seam also plays a role in deciding the seam quality [25]. Similarly, not only appearance decides the quality of the seam, the technical properties influence more on the quality of the seam [20]. Seam quality is also linked with SPI [21].

Denim is known for its durability and comfort, thus seam strength is one of the important parameters considered. The seam strength must be checked before entering the international trade as most of the consumers prefer quality. Previous studies suggest that the apparel manufacturers use appropriate seam and stitch types to ensure quality products [22]. It is observed from studies that the stitch density is directly proportional to the seam breaking load and displacement in the warp direction. The stitch density is related with the sewing thread type and affects the difference between seam breaking load before and after laundry in the weft direction [26].

Even though sewing thread contributes less than 1% by the total weight of the fabric, it is indispensable for the garment quality. Similarly, the seam type should be decided based on the design since each seam type has a significant influence on the strength of the seams [27]. When considering seam performance, seam efficiency and seam strength are essential parameters. Some of the parameters affecting seam efficiency are fabric type, weight, seam type, stitch type, stitch density, tension and strength of the thread [28].

Seam performance of garment depends on properties such as weave type, fabric thickness, weight, yarn density and stitch types. The greater the stitch density in a seam, the greater is the seam strength. Also when the number of stitches/cm is more, the strength increases. An increase in sewing thread size increases the seam strength [21]. But seam strength decreases when increasing the sewing machine speed. Thus lower speed is appreciated [29]. Different parameters are considered in measuring the seam performance such as seam efficiency, seam slippage, strength, stability, etc. Also the seam appearance and durability depends on the interrelationship between the sewing thread, needle size, stitch density, sewing machine maintenance and other conditions. It was seen that the sewing condition and the thread type play an integral role in seam quality [30].

Product quality helps the industries to survive in the competitive market. Here, not only the fabric quality is considered but also the sewing quality [31]. Sewing damage seems to be a recurring complaint from the customer's side. This can be avoided by analyzing the sewing parameters of the garment [32]. If the seam performance is poor, it affects the garment badly. Thus analyzing seam strength is essential [33].

Recycling helps in managing waste. Fletcher quotes recycling as a transition strategy [34]. Recycling helps here in collecting the pre-consumer waste from industry and then blending with other fibers so that they possess enough durability as well as aesthetics. By doing so, innovative fabrics can be developed sustainably as wastes are also prevented dumping in landfills. Fashion is considered ephemeral thus leading to fast fashion and new trends which thereby leads to a huge production of goods [35]. So, these goods must be produced in a sustainable way to reduce carbon footprints and other wastages. Reduce, reuse and recycle are the three indispensable parts of sustainability [36]. Production of garments with recycled fiber is one of the options to reduce CO_2 emission. In this study, recycled polyester is blended with cotton and elastane to make denim. This denim is compared with the basic polyester denim so that the qualities of recycled denim can be analyzed. If it is of good properties then this makes the future. The performance of the garment is indispensable to consumers. Denim known for its durability is a good choice for its performance as well as looks. Even though it is durable, the quality of seams is important and apparel industries are taking various steps in increasing the quality. Several parameters influence here.

2 Objectives

- To analyze the sustainability practices among various apparel industries.
- To develop suitable strategies regarding sustainability in apparel industries.
- To design and develop garments from recycled polyester by recycling PET bottles.
- To compare the fabric parameters between the recycled polyester denim and virgin polyester denim.
- To compare the seam strength between the recycled polyester denim and virgin polyester denim.
- To analyze and compare the comfort of the produced garments.

3 Methodology

See Fig. 1.

3.1 Selection of Sample

Recycling PET bottle waste into polyester fibers adds many benefits to the apparel industry by making a difference in reducing carbon emissions and saving energy. Thus denim blended with recycled polyester is chosen. To compare its performance,

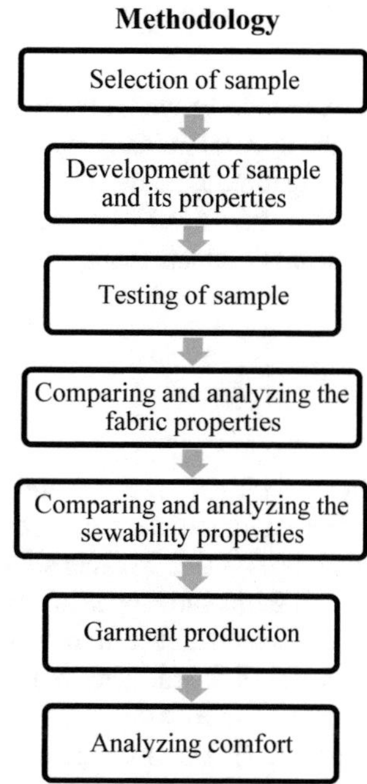

Fig.1 Flowchart of methodology

the denim blended with virgin polyester is also chosen. The fabrics are sourced from KG fabrics, Perundurai.

3.2 Development of Sample and Its Properties

In this study, two types of denim woven fabrics are taken, namely, denim made with polyester (KG 5469) blended with cotton and elastane and denim made with recycled polyester blended with cotton and elastane (ZP 5469). Properties of the fabric samples are given in Table 1.

Table 1 Fabric properties

Style No.	EPI	PPI	GSM	Tear strength (gms)		Tensile strength (*N*)	
				Warp	Weft	Warp	Weft
KG 5469	100	65	360	6208	6270	931.62	608.012
KG 5469 ZP	100	65	355	6155	6170	902.21	578.59

3.3 Testing of Sample

Commercially available sewing thread from "COATS" is used since it is widely used in the apparel industry due to its durability and stability. 25 and 50 ticket numbered sewing threads made of spun polyester are used. Three needle sizes were used namely DB*5: 100-16, 100-18 and 100-19. JUKI single needle lock stitch machine was used to seam the denim samples.

3.4 Methods

In this study, to analyze and compare the seam strength of the two types of fabric, different parameters are considered. Three different stitch densities are chosen namely 7, 8 and 9 stitches/inch. Three different seams such as superimposed seam, bound seam and lapped seam are chosen. Thus three needle sizes, at three SPI, three seams were stitched in both warp and weft way. Hence 108 samples were obtained.

Seam strength was measured according to ASTM D434-95: "Standard Test Method for Resistance to Slippage of Yarns in Woven Fabrics using a Standard seam" [37]. The Instron model 1026 Tensile Testing Machine is used to measure the seam strength which was equipped with jaws suitable for grab tests. Test specimens were cut 350 ± 3 mm by 100 ± 3 mm in both warp and weft directions. The fabric is folded 100 ± 3 mm from one end with folded parallel to the short direction of the fabric and sewn for the superimposed seam. The fabric is folded 100 ± 3 mm from one end with folded parallel to the short direction of the fabric is cut for bound and lapped seam and then sewn appropriately.

3.5 Seam Elongation

Seam elongation is the ratio of the extension of the material before stretching. The following formula helps in calculating the seam elongation.

Seam elongation percentage = ((Original Length − Extended Length)/Original Length) * 100.

3.6 Seam Efficiency

Seam efficiency is the ratio between seam strength and fabric strength. It is the ability of the fabric itself to carry a seam [30]. It is calculated with the following formula.

Seam efficiency = Seam strength/Fabric strength *100.

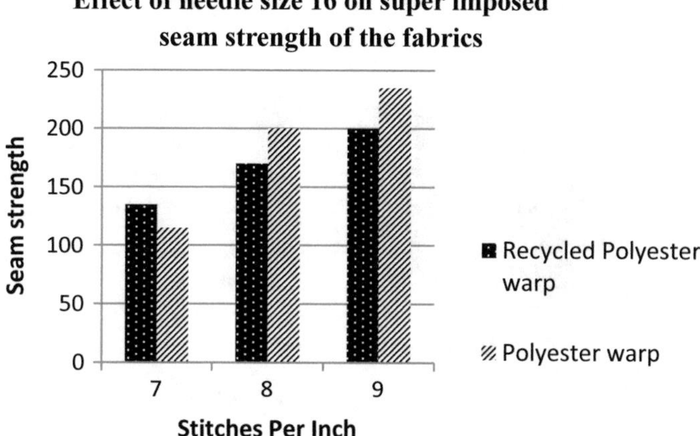

Fig. 2 Effect of needle size 16 on superimposed seam strength of the fabrics

4 Results and Discussion

4.1 Influence of Needle Size, SPI and Seam Types on Seam Strength of the Denim Fabrics

Each sample has its seam strength due to the different combinations of needle size, SPI and seam types. The results are observed via graphs as follows.

4.1.1 Effect of Needle Size 16 on Superimposed Seam Strength of the Fabrics

The superimposed seam was sewn with three SPI namely 7, 8, 9 under needle size DB*: 100-16. From Fig. 1, it was seen that the strength of polyester warp with SPI 8 and 9 is superior to the recycled polyester warp fabric of SPI 8 and 9 (Fig. 2).

4.1.2 Effect of Needle Size 18 on Superimposed Seam Strength of the Fabrics

The superimposed seam was sewn with three SPI namely 7, 8, 9 under needle size DB*: 100-18. It was observed from Fig. 2 that the strength of recycled polyester warp is superior to polyester warp in all three SPI (Fig. 3).

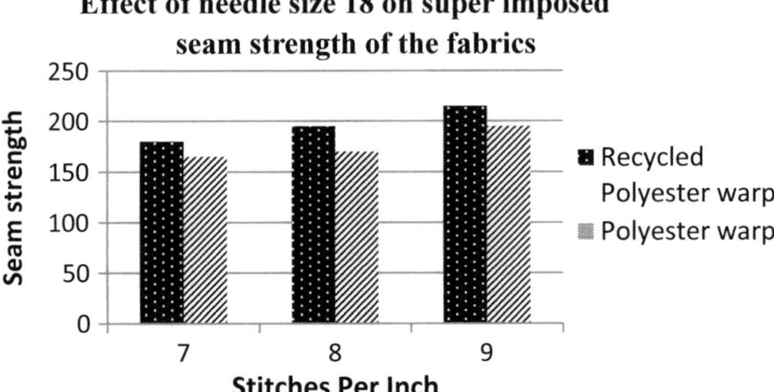

Fig. 3 Effect of needle size 18 on superimposed seam strength of the fabrics

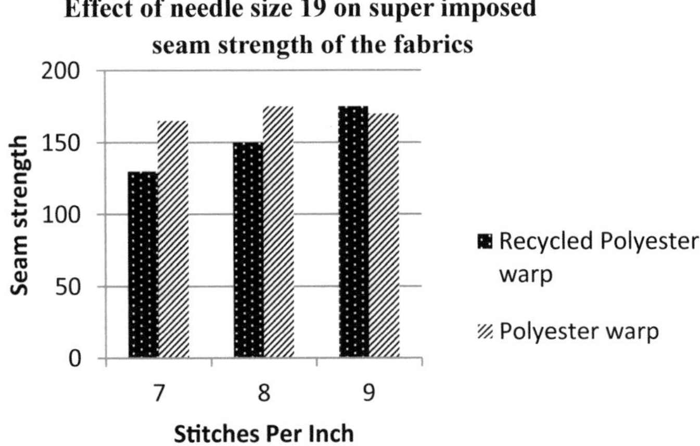

Fig. 4 Effect of needle size 19 on superimposed seam strength of the fabrics

4.1.3 Effect of Needle Size 19 on Superimposed Seam Strength of the Fabrics

The superimposed seam was sewn with three SPI namely 7, 8, 9 under needle size DB*: 100-19. From Fig. 3, it is clear that the strength of recycled polyester warp fabrics with SPI: 7 and 8 is inferior to polyester warp fabrics (Fig. 4).

4.1.4 Effect of Needle Size 16 on Bound Seam Strength of the Fabrics

A bound seam was sewn with three SPI namely 7, 8, 9 under needle size DB*: 100-16. The strength of recycled polyester warp fabrics of all three SPI is superior to polyester warp fabrics of all three SPI (Fig. 5).

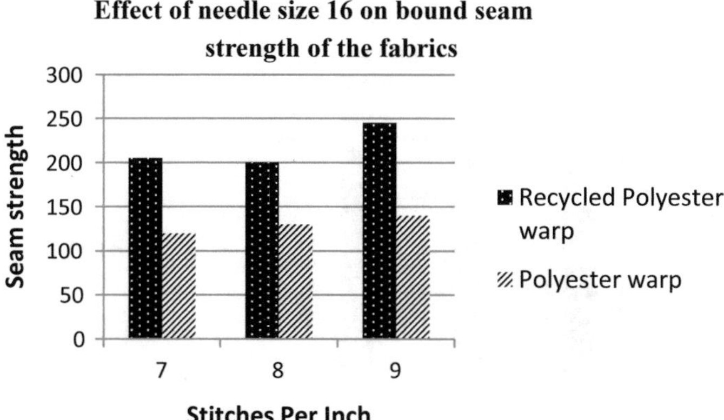

Fig. 5 Effect of needle size 16 on bound seam strength of the fabrics

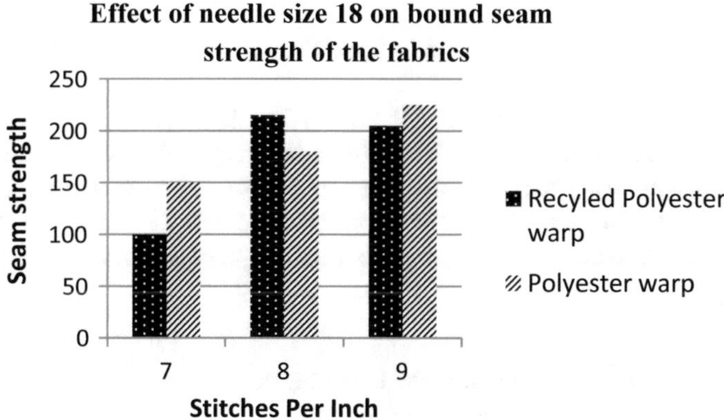

Fig. 6 Effect of needle size 18 on bound seam strength of the fabrics

4.1.5 Effect of Needle Size 18 on Bound Seam Strength of the Fabrics

A bound seam was sewn with three SPI namely 7, 8, 9 under needle size DB*: 100-18. It was observed from Fig. 5, that the strength of polyester warp with SPI 7 and 9 are superior to recycled polyester warp with SPI 7 and 9 (Fig. 6).

4.1.6 Effect of Needle Size 19 on Bound Seam Strength of the Fabrics

A bound seam was sewn with three SPI namely 7, 8, 9 under needle size DB*: 100-19. From Fig. 6, it is clear that the strength of recycled polyester fabrics of all SPI is superior to the strength of polyester warp fabrics (Fig. 7).

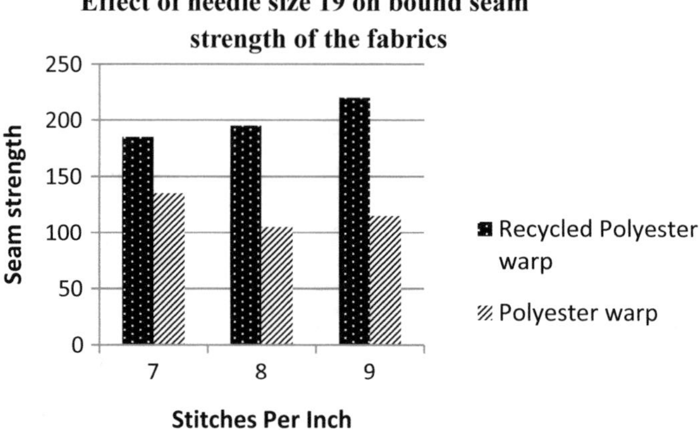

Fig. 7 Effect of needle size 19 on bound seam strength of the fabrics

4.1.7 Effect of Needle Size 16 on Lapped Seam Strength of the Fabrics

A lapped seam was sewn with three SPInamely 7, 8, 9 under needle size DB*: 100-16. It is observed from the Fig. 7 that the strength of recycled polyester and polyester warp fabrics with SPI 7 are the same. The strength of polyester warp with SPI 8 is superior to strength of recycled polyester warp fabric with SPI 8 and the strength of recycled polyester with SPI 9 is superior to the strength of polyester warp fabric with SPI 9 (Fig. 8).

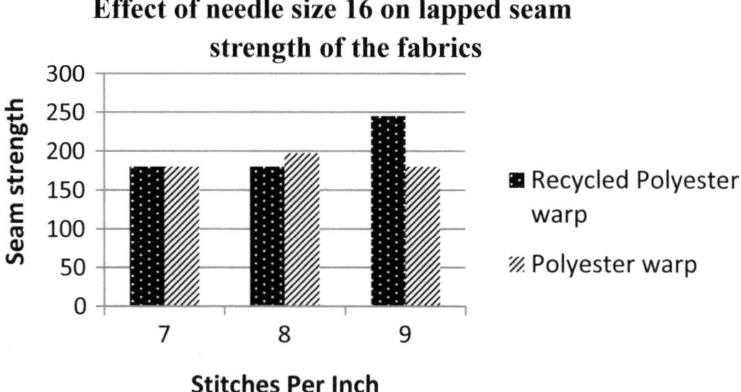

Fig. 8 Effect of needle size 16 on lapped seam strength of the fabrics

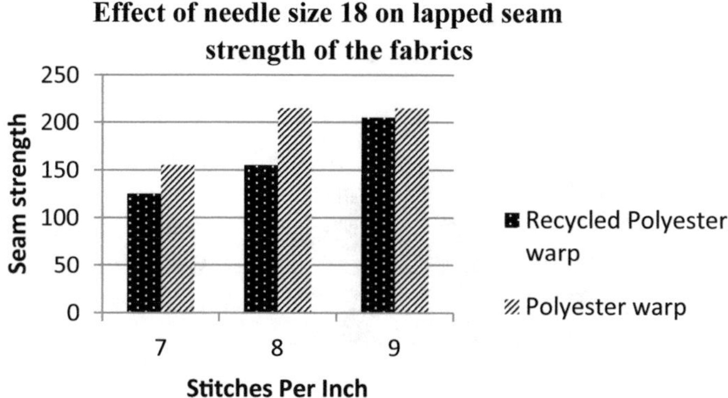

Fig. 9 Effect of needle size 18 on lapped seam strength of the fabrics

4.1.8 Effect of Needle Size 18 on Lapped Seam Strength of the Fabrics

A lapped seam was sewn with three SPI namely 7, 8, 9 under needle size DB*: 100-18. The strength of polyester warp is superior to the strength of recycled polyester warp in all SPI (Fig. 9).

4.1.9 Effect of Needle Size 19 on Lapped Seam Strength of the Fabrics

A lapped seam was sewn with three SPI namely 7, 8, 9 under needle size DB*: 100-19. The strength of polyester warp with SPI 8 and 9 is superior to the strength of recycled polyester warp with SPI 8 and 9 (Fig. 10).

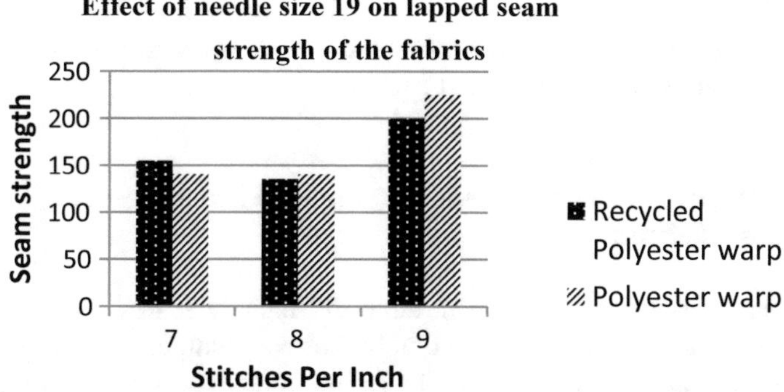

Fig. 10 Effect of needle size 19 on lapped seam strength of the fabrics

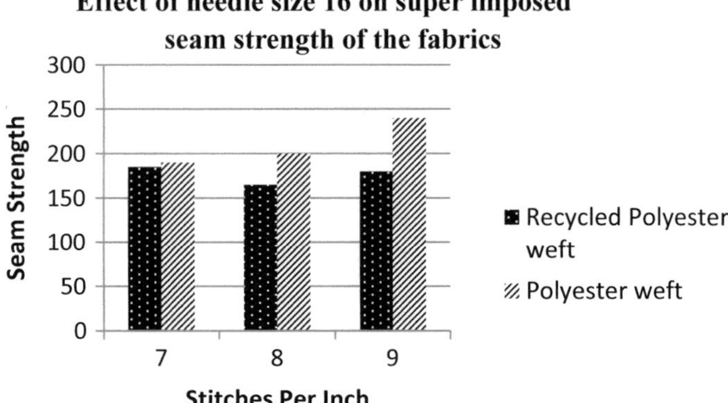

Fig. 11 Effect of needle size 16 on superimposed seam strength of the fabrics

4.1.10 Effect of Needle Size 16 on Superimposed Seam Strength of the Fabrics

A superimposed seam was sewn with three SPI namely 7, 8, 9 under needle size DB*: 100-16. It is observed from Fig. 10 that the strength of polyester weft fabrics of all SPI is superior to the strength of the recycled polyester weft fabrics (Fig. 11).

4.1.11 Effect of Needle Size 18 on Superimposed Seam Strength of the Fabrics

A superimposed seam was sewn with three SPI namely 7, 8, 9 under needle size DB*: 100-18. From Fig. 11, it was observed that the strength of recycled polyester weft with SPI 8 and 9 is superior to the strength of polyester fabric with SPI 8 and 9 whereas the strength of polyester fabric with SPI 7 is superior to the strength of recycled polyester fabric with SPI 7 (Fig. 12).

4.1.12 Effect of Needle Size 19 on Superimposed Seam Strength of the Fabrics

A superimposed seam was sewn with three SPI namely 7, 8, 9 under needle size DB*: 100-19. The strength of recycled polyester weft with SPI 7 is superior to the strength of polyester weft fabric with SPI 7. The strength of recycled polyester weft with SPI 8 is inferior to the strength of polyester weft fabric with SPI 8, whereas the strength of both recycled polyester weft and polyester weft are the same when considering SPI 9 (Fig. 13).

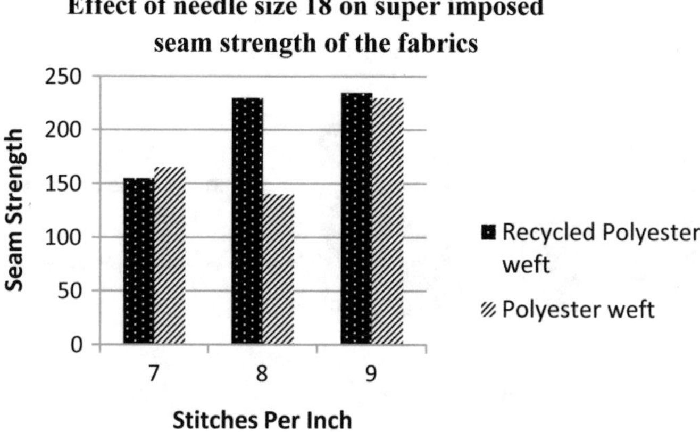

Fig. 12 Effect of needle size 18 on superimposed seam strength of the fabrics

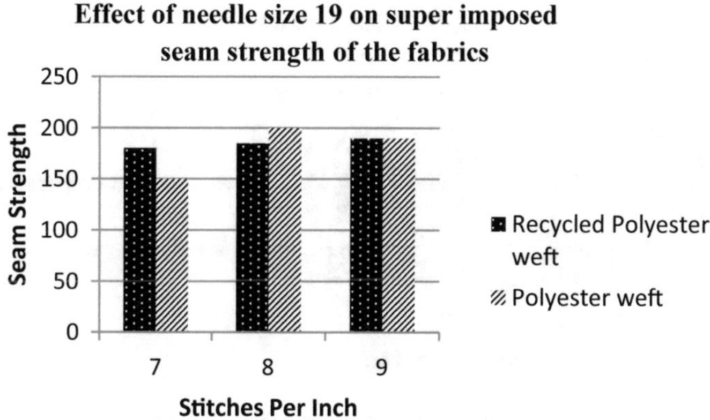

Fig. 13 Effect of needle size 19 on superimposed seam strength of the fabrics

4.1.13 Effect of Needle Size 16 on Bound Seam Strength of the Fabrics

A bound seam was sewn with three SPI namely 7, 8, 9 under needle size DB*: 100-16. The strength of recycled polyester weft fabric with all SPI is superior to the strength of polyester weft of all SPI (Fig. 14).

4.1.14 Effect of Needle Size 18 on Bound Seam Strength of the Fabrics

A bound seam was sewn with three SPI namely 7, 8, 9 under needle size DB*: 100-18. From Fig. 14, it is clear that the strength of polyester weft fabric of SPI 7 and 9 is superior to the strength of the recycled polyester weft fabric of SPI 7 and 9 (Fig. 15).

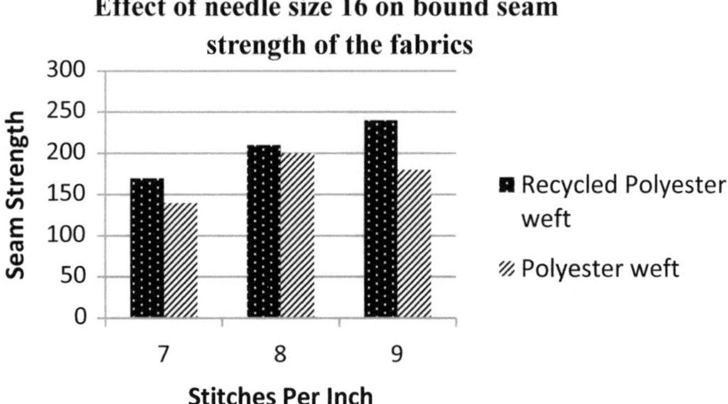

Fig. 14 Effect of needle size 16 on bound seam strength of the fabrics

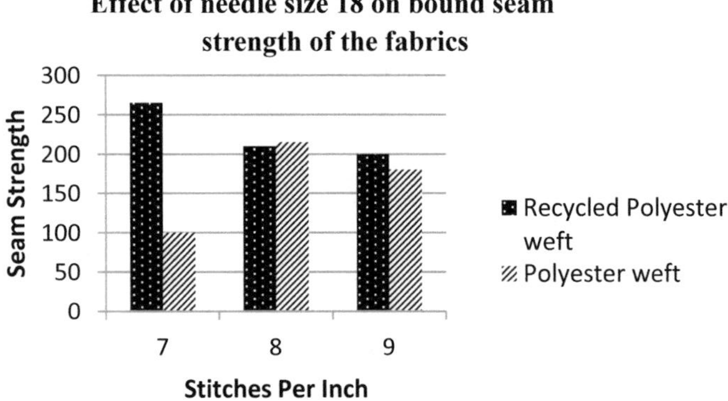

Fig. 15 Effect of needle size 18 on bound seam strength of the fabrics

4.1.15 Effect of Needle Size 19 on Bound Seam Strength of the Fabrics

A bound seam was sewn with three SPI namely 7, 8, 9 under needle size DB*: 100-19. The strength of polyester weft fabric of all SPI is superior to the strength of the recycled polyester weft fabric of all SPI (Fig. 16).

4.1.16 Effect of Needle Size 16 on Lapped Seam Strength of the Fabrics

A lapped seam was sewn with three SPI namely 7, 8, 9 under needle size DB*: 100-16. From Fig. 16, it is clear that the strength of recycled polyester weft fabric with SPI 7 and 9 is superior to the strength of polyester weft fabric with SPI 7 and 9 (Fig. 17).

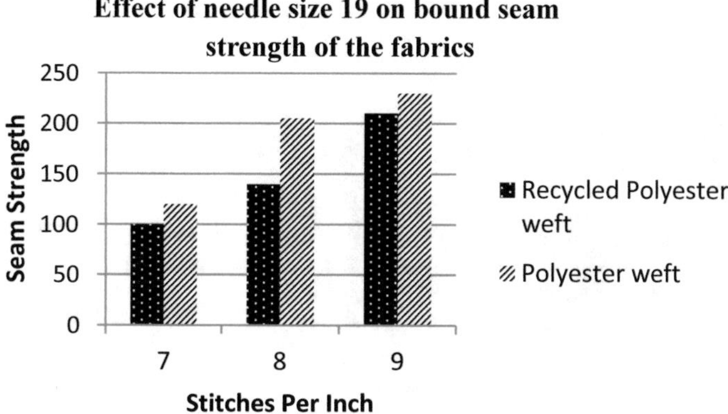

Fig. 16 Effect of needle size 19 on bound seam strength of the fabrics

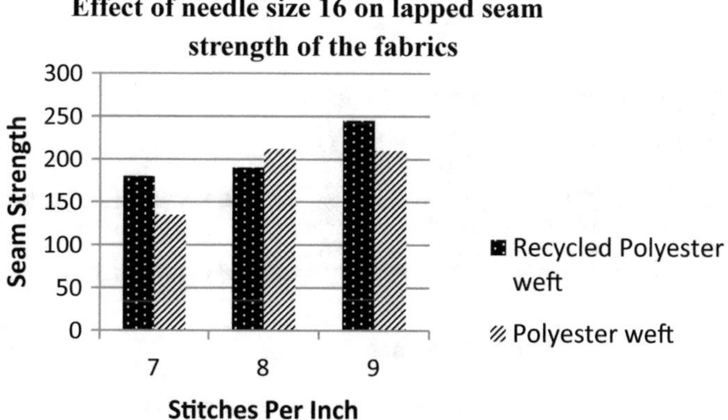

Fig. 17 Effect of needle size 16 on lapped seam strength of the fabrics

4.1.17 Effect of Needle Size 18 on Lapped Seam Strength of the Fabrics

A lapped seam was sewn with three SPI namely 7, 8, 9 under needle size DB*: 100-18. The strength of recycled polyester weft fabrics with all SPI is superior to the strength of polyester weft fabrics with all SPI (Fig. 18).

4.1.18 Effect of Needle Size 19 on Lapped Seam Strength of the Fabrics

A lapped seam was sewn with three SPI namely 7, 8, 9 under needle size DB*: 100-19. From Fig. 18, it is observed that the strength of recycled polyester weft fabrics with SPI 8 and 9 is inferior to the strength of polyester weft fabrics of SPI 8 and 9 (Fig. 19).

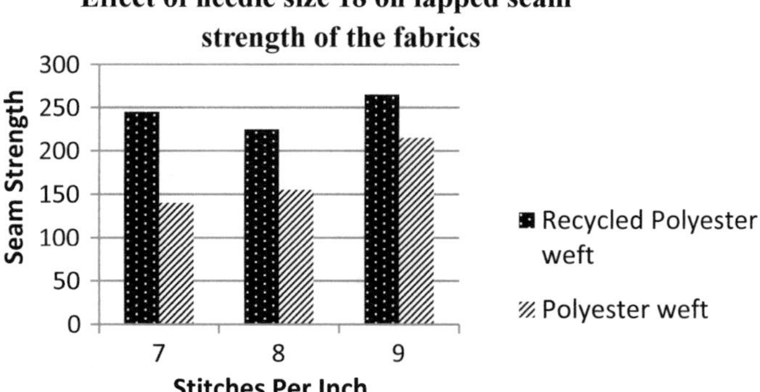

Fig. 18 Effect of needle size 18 on lapped seam strength of the fabrics

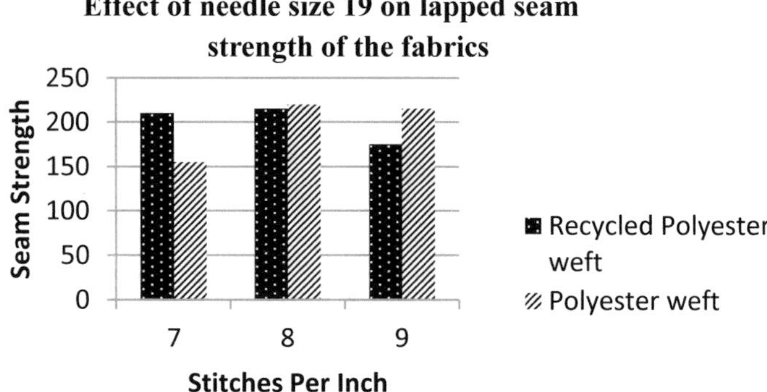

Fig. 19 Effect of needle size 19 on the lapped seam strength of the fabrics

4.2 *Effect of SPI, Seam Type and Needle Size on Seam Efficiency of Recycled Polyester and Polyester Denim Fabrics*

Considering the seam types, needle size and SPI, the seam efficiency is calculated. Among samples stitched with Superimposed seam, the sample with needle size 18 and SPI 9 has the highest score in seam efficiency in ZP5469 warp; the sample with needle size 16 and SPI 9 shows high seam efficiency in KG5469 warp; the sample with needle size 18 and SPI 9 has the highest score in seam efficiency in ZP5469 weft; the sample with needle size 16 and SPI 9 shows high seam efficiency in KG5469 weft.

Among samples stitched with Bound seam, the sample with needle size 16 and SPI 9 has the highest score in seam efficiency in ZP5469 warp; the sample with needle size 18 and SPI 9 shows high seam efficiency in KG5469 warp; the sample with needle size 18 and SPI 7 has the highest score in seam efficiency in ZP5469 weft; the sample with needle size 19 and SPI 9 shows high seam efficiency in KG5469 weft.

Among samples stitched with Lapped seam, the sample with needle size 16 and SPI 9 has the highest score in seam efficiency in ZP5469 warp; the sample with needle size 19 and SPI 9 shows high seam efficiency in KG5469 warp; the sample with needle size 18 and SPI 9 has the highest score in seam efficiency in ZP5469 weft; the sample with needle size 19 and SPI 9 shows high seam efficiency in KG5469 weft (Table 2).

Table 2 Seam efficiency

Seam type	Needle size	SPI	ZP 5469 warp	KG 5469 warp	ZP 5469 weft	KG 5469 weft
SI seam	16	7	66.56	54.91	142.22	139.00
		8	83.81	95.49	126.85	146.31
		9	98.60	112.20	138.38	175.58
	18	7	88.74	78.78	119.16	120.71
		8	96.14	81.17	176.82	102.42
		9	106.00	93.10	180.66	168.26
	19	7	64.09	78.78	138.38	109.73
		8	73.95	83.55	142.22	146.31
		9	86.28	81.17	146.07	139.00
Bound seam	16	7	101.07	57.29	130.69	102.42
		8	98.60	62.07	161.44	146.31
		9	120.79	66.84	184.50	131.68
	18	7	49.30	71.62	203.72	73.16
		8	106.00	85.94	161.44	157.29
		9	101.07	107.43	153.75	131.68
	19	7	91.21	64.46	76.88	87.79
		8	96.14	50.13	107.63	149.97
		9	108.46	54.91	161.44	168.26
Lapped seam	16	7	88.74	85.94	138.38	98.76
		8	88.74	94.06	146.07	155.09
		9	120.79	85.94	188.35	153.63
	18	7	61.63	74.00	188.35	102.42
		8	76.42	102.65	172.97	113.39
		9	101.07	102.65	203.72	157.29
	19	7	76.42	66.84	161.44	113.39
		8	66.56	66.84	165.28	160.94
		9	98.60	107.43	134.53	157.29

4.3 Effect of SPI, Seam Type and Needle Size on Seam Elongation of Recycled Polyester and Polyester Denim Fabrics

The seam elongation percentage was calculated for both recycled polyester and polyester fabrics for all three types of seams considering three needle sizes and SPI The results were shown in Table 3.

Table 3 Seam elongation percentage

Seam type	Needle size	SPI	Recycled polyester warp	Polyester warp	Recycled polyester weft	Polyester weft
SI seam	**16**	7	2.4	2.8	2.1	2
		8	2.5	3.1	2.2	2.2
		9	2.4	3.4	2.1	2.3
	18	7	2.5	2.1	1.9	2.1
		8	2.7	3	3.3	2.5
		9	2.8	2.9	2.3	2.2
	19	7	2.1	2.6	1.5	2.7
		8	2.8	2.9	2.1	1.6
		9	2.7	2.5	1.9	2.1
Bound seam	**16**	7	3.2	2.9	1.6	2.5
		8	2.9	2.9	2.3	2.8
		9	3.3	2.9	2.3	2.8
	18	7	3.2	3.4	2.6	2.6
		8	2.9	2.5	2.4	2.2
		9	3.1	3.1	2.3	3
	19	7	3	2.3	3	2.9
		8	2.8	2.8	3	2.9
		9	3.3	2.7	2.6	2.2
Lapped seam	**16**	7	3.1	2.3	2.1	1.9
		8	2.7	2.8	1.9	2.4
		9	3.1	2.4	2.3	2.9
	18	7	2.6	2.1	2.1	2.1
		8	2.9	2.6	2.2	2.1
		9	2.7	2.8	2.9	2
	19	7	2.6	2.3	2	2.3
		8	2.3	2.7	2.1	2.1
		9	2.8	2.9	2	2

Table 4 Shrinkage results

Particulars	ZP 5469 (recycled polyester)	KG 5469 (polyester)
Warp	3	4
Weft	15	16
Skew	2.4	1
Shrinkage	3*15	4*16

Table 5 Colorfastness

Parameters	ZP 5469 (recycled polyester)	KG 5469 (polyester)
Rubbing fastness	Dry—3	Dry—3
	Wet—1.5	Wet—1.5
Washing fastness		
Acetate	4	4
Cotton	4.5	4.5
Nylon	4	4
Polyester	4	4
Acrylic	4.5	4.5
Wool	4.5	4.5

4.4 Shrinkage

Both polyester and recycled polyester denim shows similar results. Usually the skewness should lie within 3, thus both fabrics are good. Lesser the shrinkage is good for fabric, thus recycled polyester comparatively shows good results (Table 4).

4.5 Colorfastness

The colorfastness of both the fabrics was analyzed based on rubbing and washing. Both the fabrics gave the same results (Table 5).

4.6 Analysis of Comfort in the Produced Garments

See Figs. 20, 21, 22, and 23.

Two styles viz., a Capri and a Trouser is stitched in both recycled polyester and polyester denim fabric. Size 38 is stitched and both styles were enzyme washed as this wash is comparatively sustainable. Subjective analysis is done based on a scale of

Fig. 20 Trouser made with recycled polyester

1, 2, 3, 4, 5 and 6 which represent the most satisfactory, very satisfactory, satisfactory, unsatisfactory, very unsatisfactory and most unsatisfactory, respectively (Table 6).

5 Conclusion

Denim is an evergreen fabric. Since an apparel industry is one of the most polluting industries, it is the responsibility of each industry to incorporate sustainable strategies. In this project work, the recycled polyester denim fabric is compared with Polyester denim fabric to check whether there are any major differences present between both the fabrics. The results prove that there are minor differences only and thus, recycled polyester can be used successfully. The seam strength results show that the recycled polyester possesses good strength. The comfort of both fabrics as garments was also similar. Thus replacing virgin polyester with recycled polyester can be successfully incorporated by the apparel industry striving toward sustainability.

Fig. 21 Trouser made with virgin polyester

Fig. 22 Capri made with recycled polyester

Fig. 23 Capri made with virgin polyester

Table 6 Analysis of comfort

S.No.	Questions	Rating
1	Is the garment made out of recycled polyester comfortable?	3
2	Is the garment of good ease while working or in motion?	2
3	Is Capri style more comfortable than trousers?	2

References

1. Vijayalakshmi D (2015) A study on the anti-bacterial property of essential oils and metal oxides in denim garments. Int J Text Fash Technol (IJTFT) 5(1):49–54
2. Gokarneshan N, (2018) Advances in denim research. Research and Development in Material Science 3(1)
3. Marmarah A (2017) New knitted fabric concepts for denim products. 17th World Textile Conference AUTEX 2017—Textiles—Shaping the Future
4. Mashiur Rahman Khan Md. (2011) Effect of bleach wash on the physical and mechanical properties of denim garments. Proceedings of the International Conference on Mechanical Engineering 2011 (ICME2011)
5. Kan CW (2013) Prediction of color properties of cellulase-treated 100% cotton denim fabric. J Text 2013:1–10
6. Muthu SS (ed) (2014) Roadmap to sustainable textiles and clothing: environmental and social aspects of textiles and clothing supply chain. Springer, Singapore
7. Dascalu T (2000) Removal of the indigo color by laser beam denim interaction. Opt Lasers Eng 34(3):179–189
8. Juciene M (2014) The effect of laser technological parameters on the color and structure of denim fabric. Text Res J 84(6):662–670
9. Hirscher AL (2013) Fashion activism evaluation and application of fashion activism strategies to ease transition towards sustainable consumption behaviour. Res J Text Appar 17(1):23
10. Chopra N, Singh M. Mapping the environmental sustainability practices in a garment manufacturing unit: a case study analysis to identify inhibitors. Int J Text Fash Technol 9(4):11–18
11. Elahi S, Hosen MD, Islam M, Hasan Z, Helal MM, Al MS, Rakin S (2019) Analysis of physical & chemical properties of cotton-jute blended denim after a sustainable (industrial stone enzyme) wash. J Textile Sci & Fashion Tech 3(2):1–8
12. Chatterjee KN, Sharma D, Pal H (2020) Quality aspects of sustainable handloom denim fabrics made of hand spun and machine spun cotton yarn. J Text Appar Technol Manag 11(2)
13. Al-Sabaeei A, Napiah M, Sutanto M, Alaloul W (2019) Effects of waste denim fibre (WDF) on the physical and rheological properties of bitumen. In: IOP conference series: materials science and engineering, vol 527, no 1. IOP Publishing, p 012047
14. Sandin G, Peters GM (2018) Environmental impact of textile reuse and recycling—a review. J Clean Prod 184:353–365
15. Vadicherla T, Saravanan D (2017) Thermal comfort properties of single jersey fabrics made from recycled polyester and cotton blended yarns. Indian J Fibre & Text Res (IJFTR) 42:318–324
16. Sarıoğlu E (2017) Ecological approaches in textile sector: the effect of r-pet blend ratio on ring spun yarn tenacity. Period Eng Nat Sci 5(2):176–180
17. McCullough H, Sun D (2019) An investigation into the performance viability of recycled polyester from recycled polyethylene terephthalate (R-PET). J Text Sci Fash Technol 2(4):1–8
18. Bhatia D, Sharma A, Malhotra U (2014) Recycled fibers: an overview. International Journal of Fiber and Textile Research 4(4):77–82
19. Mert ATEŞ, Gürarda A, Çeven EK (2019) Investigation of seam performance of chain stitch and lockstitch used in denim trousers. Tekstil ve Mühendis 26(115):263–270

20. Islam MM, Saha PK, Islam MN, Rana MM, Hasan MA (2019) Impact of different seam types on seam strength. Glob J Res Eng 19(4)
21. Jaber M, Islam MM (2019) Investigating seam strength & seam performance with different SPI on different fabrics. IOSR J Polym Text Eng (IOSR-JPTE) 6(2):32–41
22. Seetharam G, Nagarajan L (2014) Evaluation of sewing performance of plain twill and satin fabrics based on seam slippage seam strength and seam efficiency. IOSR J Polym Text Eng (IOSR-JPTE) 1(3):9–21
23. Maaroul MA (2015) Effect of the seam efficiency and pukering on denim sewability. J Basic Appl Sci Res 5(10):24–32
24. Bharani M, Gowda RM (2012) Characterization of seam strength and seam slippage of PC blend fabric with plain woven structure and finish. Res J Recent Sci. ISSN: 2277-2502.
25. Trivedi V, Singh GP, Khanna S (2018) Effect of sewing process on tensile properties of sewing threads in denim garment. Journal of the textile association, 260
26. Elsheikh KM, Shawky M, Darwish HM, Elsamea EA (2018) Prediction of seam performance of light weight woven fabrics. Int J Eng Tech Res 8(6)
27. Chen D, Cheng P (2019) Investigation of factors affecting the seam slippage of garments. Text Res J 89(21–22):4756–4765
28. Iftikhar F, Hussain T, Malik MH, Ali Z, Nazir A (2018) Fabric structural parameters effect on seam efficiency-effect of woven fabric structural parameters on seam efficiency. J Textile Sci Eng 8(358):2
29. Debes RMKA, Mazari AA (2019) Optimizing sewing speed for better seam quality of denim fabric. Vlakna a Textil (2)
30. Malek S, Kheder F, Jaouachi B, Cheikhrouhou M (2019) Influence of denim fabrics properties and sewing parameters upon the seam slippage and seam quality prediction. J Text Appar Technol Manag 11(1)
31. Bakici GG, Kadem FD (2018) An experimental study on sewability properties of 100% cotton denim fabrics. Tekstil ve Konfeksiyon 28(2):170–176
32. Padhye R, Nayak R (2010) Sewing performance of stretch denim. J Text Appar Technol Manag 6(3):1–9
33. Rajput B, Kakde M, Gulhane S, Mohite S, Raichurkar PP (2018) Effect of sewing parameters on seam strength and seam efficiency. Trends Text Eng Fash Technol 4(1):1–5
34. Aakko M, Koskennurmi-Sivonen R (2013) Designing sustainable fashion: possibilities and challenges. Res J Text Appar 17(1):13
35. Sanches RA, Takamune KM, Guimarães BMG, Seawright RA, Karam Jr D, Marcicano JPP, Duarte, A, Dedini FG (2015) Comparative study of the characteristics of knitted fabrics produced from recycled fibres employing the Chauvenet criterion, factorial design and statistical analysis. Fibres Text East Eur 23(4):19–24
36. Yamunanagar HI (2018) Green fashion: need of the hour for sustainable development (a review). Res Rev Int J Multidiscip 11(4)
37. ASTM D434-95 (1995) Standard test method for resistance to slippage of yarns in woven fabrics using a standard seam (withdrawn 2004). ASTM International, West Conshohocken, PA, www.astm.org

A Look Back at Zero-Waste Fashion Across the Centuries

B. Subathra and D. Vijayalakshmi

Abstract Sustainable fashion is an endeavor that draws together sustainable development and fashion. In recent years, the fashion industry has received abundant criticism over its limited consideration of social and environmental welfare issues on the global public agenda. The concept of zero waste design includes many different approaches which all aim to eliminate fabric waste. Although its name is new, the idea is much older: for example the traditional Japanese kimonos or Indian sarees both make use of one complete piece of a fabric without wasting any of it. Most of the centuries, could find the trace of traditional clothing were used as multipurpose wear. A rectangular piece of cloth was worn as garment, used at bedsits, cradle, as shawl or head-gear at different climates. The traditional rectangular piece of cloth, found as unisex wear, was used by both men and women. At each century could see the zero-waste clothing's over different countries like Egypt, Greece, Rome, Japan, and Korea. Even in the twenty-first century without emphasizing the concept, few zero-waste clothing are used by the Indian people. Zero-waste approach means designing without wasting fabric. While pattern making, the awareness of parameters and space creates the endless possibilities of zero-waste techniques. A key aim of this study is to analyze the traditional clothing of various countries over the centuries and to provide the industry with findings that can be applied to design garments, which can also manufactured and to suggest designers, to what extent zero-waste approach is feasible and desirable within contemporary fashion industry.

Keywords Contemporary · Sustainable fashion · Zero-waste design approach

B. Subathra (✉) · D. Vijayalakshmi
Department of Apparel and Fashion Design, PSG College of Technology, Coimbatore, India
e-mail: bsa.afd@psgtech.ac.in

D. Vijayalakshmi
e-mail: hod.afd@psgtech.ac.in

S. S. Muthu (ed.), *Sustainable Approaches in Textiles and Fashion*, Sustainable Textiles: Production, Processing, Manufacturing & Chemistry,
https://doi.org/10.1007/978-981-19-0530-8_5

1 Introduction

Fashion is defined as, a trend that occurs at a specific era and area: It influences the people to think that what makes a sustainable lifestyle, such as whether fast fashion is sustainable, slow fashion is sustainable, or whether it should be more inclusive or exclusive. Fashion, resides at the nexus of desire and discard, between the glistening yearning to live and the sweltering heat of death. When it comes to the clothing themselves, their durability is decided by their use and "metabolism"—certain garments are made to endure a long time, such as outdoor wear and winter jackets, while others, such as a party top, are designed to be worn quickly.

Based on the clothing need, some can be manufactured to be either more durable, while others should be compostable or recyclable to decay more quickly. The aging of some garments acquires a patina and enchantment comparable to the wonder, fascination, and splendor of ruins, whereas the discarded rags of last season are an eyesore and a nuisance; the first denotes a majesty of taste and the second, waste.

A fundamental reason for the current unsustainable status of the fashion system is the rapid rush of new products onto the market, or fast fashion. Fast fashion is a catch-all word for low-cost, easily available, and on-trend clothing acquired through global supply chains and sold by well-known brands. Every year, the fashion industry spends three trillion dollars. It accounts for 2% of worldwide GDP, which is defined as the total value of all finished products and services produced inside a country's borders. The majority of the three trillion dollars comes from fast fashion.

The "rapid" part of consuming is primarily a problem for the environment when done on a large scale. Because fast conspicuous spending was mostly limited to the wealthy, its global influence was overlooked or dismissed. As a result, purchasing haute couture quickly is not seen as a problem, but rather as praise, whereas purchasing fast fashion by those with lower means is seen as unethical and inappropriate. Fast fashion has become the standard in an industry that is constantly releasing new collections and products: a high-quality garment does not always imply a lower rate of consumption and waste. Fashion trends are speeding up, exacerbating the situation. The demand for apparel has increased as a result of the short duration of micro-trends.

2 Fast Fashion

The fast-fashion business harms the environment and is unethical in terms of production. To keep up with shifting fashion trends, clothing is made in dangerous methods due to no other options to satisfy customers. Synthetic fibers are used instead of natural fibers in "fast" apparel. Fossil fuels are used to make these fibers. This is expected to account for 66 percent of our apparel. Landfills are quickly filling due to the high cost of such clothes and the regularity with which they are purchased. Fashion products account for over 20% of global trash, with more than 60% of clothes manufactured each year ending up in landfills as consumer waste. Low salaries and exploitative working conditions are typical in factories that create fast clothes, and these factories are often exploitative.

3 Slow Fashion

A slow fashion movement, based on the concepts of slow food, can be regarded as a counter-force to fast fashion. In the same way that emotional, ecological and ethical elements are prized over uniformity and dull convenience, sustainability aligns with "slow fashion" ideas. It is necessary to change infrastructure and reduce commodities throughput. Slow fashion, in general, isn't about business as usual, nor is it solely about design. The slow fashion movement isn't typical fashion trend, and it's not limited to traditional styles. Slow fashion is about creating unique, personalized goods while respecting our living situations, biological diversity, cultural identity, and limited resources.

Slow Fashion emerged as a result of a need. Kate Fletcher coined the name following the slow food movement. Similar to the slow food movement, she saw a need for a slower pace in the fashion industry.

Fashion's growth is concerned with mass production and worldwide style is challenged by slow fashion. It becomes a protector of diversity and shifts power dynamics between fashion designers and consumers. It is resulting in the formation of new relationships and trust that can only be achieved on a smaller scale. The design process promotes an understanding of how the process affects resource flows, employees, communities, and ecosystems.

4 Sustainable Fashion

Few decades before, there was a culture that people used very limited garments and accessories. People wore comparatively few clothing before the start of mass manufacture. Bespoke outfits were created for high-status individuals, whereas low-status individuals created their attire by their own hands. Repairing and remodeling procedures allowed the clothing to be preserved and utilized for longer periods. In a family, an article of clothing created for the elder son is worn by him for a few years before, being worn by the family's youngest son at the same age. These are long-lasting items. Consumers bought more, because the ready-to-wear fashion culture standardized sizes and cost reductions due to large production, even if the fit was poor. Simultaneously, globalization of manufacturing, greater competitiveness, and consumer demand have expedited the fashion cycle, resulting in a culture of rapid and throwaway fashion.

5 Sustainable Fashion in History

The concept of zero-waste clothing technique is not a new idea to get confused to adopt in bulk production. The people in the old centuries used minimum resources from Mother Nature. Ancestors tried to safeguard nature. Multipurpose clothing was adopted in those periods. A long cloth used as a shawl during winter was converted to a pouch to hold the grains, then used as a covering cloth of furniture, it was later used as a bed sheet during night time.

5.1 Before Common Era

5.1.1 Egypt

Egypt is located in the Northeastern corner of Africa. Nile River is the reason for its prosperity. Abundant cultivation of flax plant is found here and Linen is the fabric made from it. Egypt's climate is hot and sunny where linen gives perfect complement to this climatic condition. The hot climate made the people want to wear an article of lightweight clothing. One of the most appealing properties of linen was its thinness. Thin and sheer linen could be made to be transparent.

The loincloth was a very small garment that covers the private part, which was worn by the working men. Later years, the length of the loincloth reached till mid-thigh and the loin skirt also evolved. The main form of dress of kings in the olden days was Schenti, a small kilt tied around the waist falls till knee cap. Kalasiris was a long linen dress, made with very fine, thin yarn. This transparent dress was used by wealthy women; heavy coarser kalasiris looks very thick, and that was the clothing of poorer women. Cleanliness was considered more important in Egyptian culture than decorations.

In the later periods, the bell-shaped tunics, which were fitting close to the shoulders and flaring toward the ends were in fashion. The women's attire was more or less similar to that of men. It consisted of a simple the narrow straight garment which reached from the breast to the ankle and was held up by two straps on the shoulders (Figs. 1 and 2).

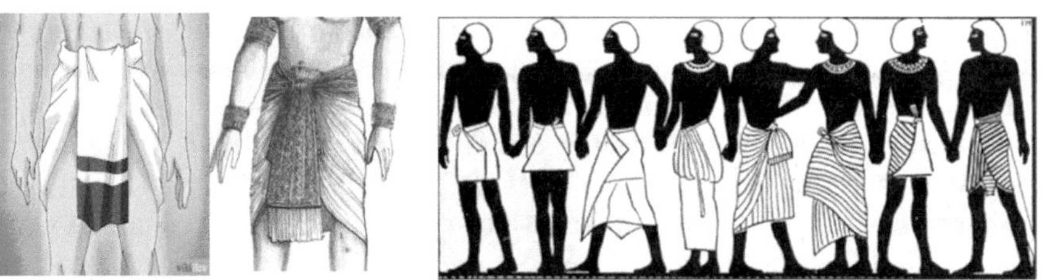

Fig. 1 Labours Loin cloth and king's Schenti and its various draping styles

Fig. 2 Draping styles of Kalasiris & bell shaped and narrow tunic

5.1.2 Greek

Greece is located in south-east Europe; it has a famous tourist place of Rome and Athens. Greek shares its boundaries with Libya and Egypt, Italy, and Cyprus. Ancient Greece is well-known for its rich heritage civilization and simplicity of the dress. Chiton and Himation were the traditional unisex garments that are always associated with the life of Greek. The garment which resembles the Chiton was used by women named peplos. It was a large square piece of fabric folded near the bust region and tied. This folded part was called Apoptygma (Figs. 3 and 4).

The chaste and refined simplicity of Greek dress is a legacy of ancient Greek civilization to posterity. Throughout Greek history, men and women wore the Chiton and the Himation, both associated with Greek culture. Nowadays, those are frequently called tunic and the mantle. The outer garment worn by the Greeks was the Himation. It was an oblong piece of Linen fabric measuring about fifteen feet long and six feet wide. It was wrapped over the body so intricately that no claps were required to keep it in place.

Ancient Greek women wore peplos as an outer garment. This was a large square-shaped piece of cloth. Using a double layer of cloth to cover the chest is called

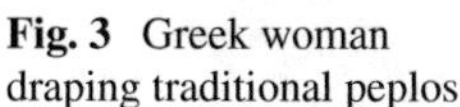

Fig. 3 Greek woman draping traditional peplos

Fig. 4 Draping method of Ionic peplos and Doric peplos

Apoptygma. Wrapping the cloth around the body and fastening it with breeches over each shoulder.

The ionic or laconic *peplos* was a rectangular piece of fabric wound around the body and the edges of the fabric were kept at the right side shoulder, which makes an opening at the right side of the body. The left part is covered with fabric and the right side fabric edges are tied as a knot. The doric peplos was a kind of garment sewn on the edges and made as tubular fabric (Fig. 5).

Peplos were usually worn with a belt. Belts can be worn either over or under Apoptygma. The Chiton was a form of the tunic which was doubtless of Asiatic origin. The earliest form of Greek chiton was called the Peplos. This was in use during the period 1200BCE–600BCE. Ancient Greece wore the Doric chiton as one of the most popular garments for men and women.

Fig. 5 Draping styles of Chitin

5.2 *Common Era*

5.2.1 1–5th Century

Rome

Ancient Rome is described as a Roman civilization, which developed from a small town situated on the Tiber River in central Italian Empire. The arts and architecture of the Roman Empire are parallel to the history of Rome. Many of their ideas came from the Greeks, but as the empire expanded, people from many different cultures, climates, and religions were integrated into the empire. Greece and Rome are nearby, both countries are separated by the Ionian arm of the Mediterranean Sea. It is not surprising that they were well aware of each other. The tradition and culture were spread one from the other.

The expansion of the empire led to wider trade opportunities. As a result, a greater variety of elegant fabrics became available. The wealthy could access high-quality embroidered edging and fringing on cotton and silk from India and East Asia. History reveals that Elagabalus was the first Roman king, who used silk clothing.

The toga, a wraparound robe worn by ancient Romans, was their favored choice. The Toga, which was worn by both men and women over the long tunic or Stola, which was akin to the Greek Chiton, was considered the Romans' national clothing. Wrapping a huge part of cloth material wound around the male body like a cloak is a common in that era. A toga functioned similarly to the Greek Himation. The peculiar draping method of the toga gained a special distinction under the empire.

Togas were draped across a figure starting with the left foot. Bypassing the straight edge underneath the right arm, over the shoulders, and across the back, the point was reached. Over the left shoulder, it was carried across the chest, leaving the remaining portion hanging from the back. If the ends were long they were shortened by tying knots, however, sometimes as a sign of great dignity, the ends were allowed to sweep the ground.

Heavy woolen fabric in the form of a half circle was also used as a toga with the unique draping style. The size of the fabric is 18 feet, which is approximately 5.5 m. To drape the cloth, 5 feet of the straight edge from the floor level were positioned in the center front of the body. The material was passed over the left shoulder, twisted around the back, right arm, and then passed over the left shoulder once more. As a result, the right arm became free.

The color of the material and the surface embroidered rose was famous in clothing. The basic clothing style is termed Chiton or Tunica, which was famous for its color of the material and rose-shaped surface embroidery. The color of the garment distinguishes the classes of society. Royal people found the colors purple and gold. Men in high rank wore a striped purple toga (Fig. 6).

Purple silk toga with golden embroidery-badge was worn by the empire families. Blue for the philosopher, Black for theology, Green toga for medicine people, White for soothsayer without ornaments. The peasantry was allowed to use only one color,

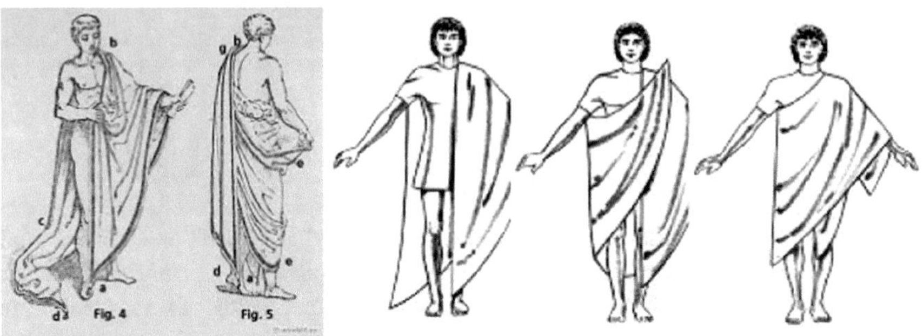

Fig. 6 Draping method of traditional Toga

Fig. 7 Multiple layer clothing style as new fashion

the officers two, the commanders three, and for the royal household seven colors were permissible.

Palla was distinctly a woman's garment never adopted by men. It was worn over a long tunic or stola. The stola resembled the Greek Chiton and fell in numerous folds about the feet. Originally it had sleeves reaching the elbow and later extended to the wrist. The lower edges of the palla were often trimmed with border and fringes. Roman women wore several tunics, one over the other in different colors, to express their luxurious taste. Mamillare was a breast band used by women of Rome (Figs. 7, 8, 9, 10, and 11).

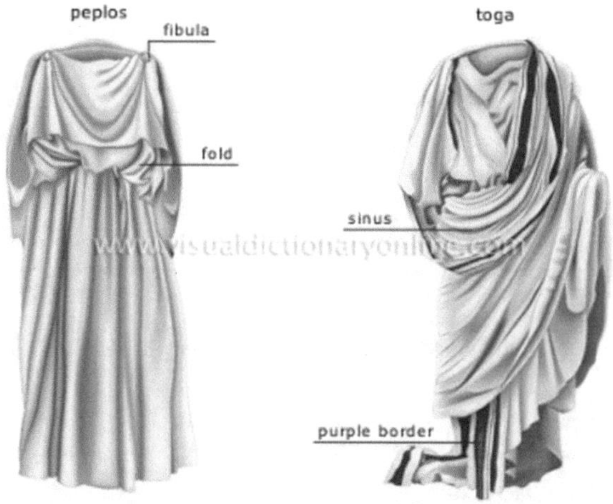

Fig. 8 Male & female clothing of Rome1

Fig. 9 Roman's Dalmatica original Dalmatica kept in museum

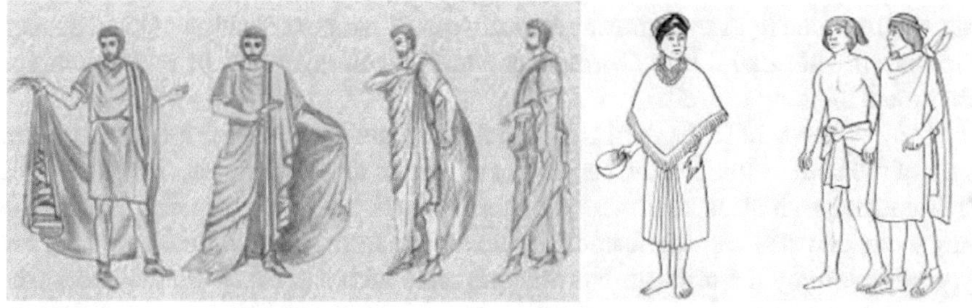

Fig. 10 Costume of Philosopher, Roman costumes of common people-Dalmatica, Toga, Mantle

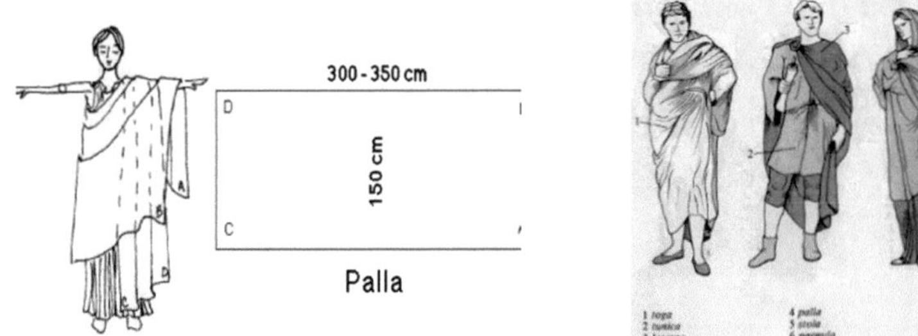

Fig. 11 Traditional Roman Women's clothing & traditional Men's clothing

Fig. 12 Costumes of Middle East countries

5.2.2 6–10th Century

Middle East Countries

The Middle East had an incredibly uniform costume throughout its history. A large part of this is due to its evolution as a weatherproof garment, used as a shield against various climatic conditions. Numerous Muslim countries accept and encourage traditional clothing as well.

Many of the Middle Eastern traditional garments are free size fitting and cover most of the body. Although the garments are used in many countries, they are named differently in each place. A similar variation can be found in the materials from which they were, and still are, fabricated. Textiles made from linen, cotton, and wool are typically worn by the affluent, but rich garments with silk bases are the exception. Several famous materials come from here, such as baldachin, a richly decorated fabric, gold yarns were interlaced with silk threads (Fig. 12).

Fig. 13 Mantle S

Arabian Countries

Many of the traditional garments of this region originally originated in ancient cultures, particularly in Persia (Iran), India, Mongolia, and Asian Russia. This is the case with the caftan. The garment is an open, coat-like garment called a candys or kandys in ancient Persia. This style of clothing was also common in Mongolia and China, and it eventually extended westward to become, in the late Ottoman Empire, the fashionable dolman.

The typical Middle Eastern clothing spread as Arab civilizations and people spread throughout the Middle East. At the time, the most typical apparel consisted of a loose shirt, chemise, or robe, a draped cloak, broad, baggy trousers, and a turban or head cloth. Street vendors still offer these items in Egypt, Istanbul, and Damascus.

A simple, long shirt, tunic, or chemise, often with long sleeves, was the basic garment for both sexes. There was a robe or mantle over this which was worn by men of varying types. Ancient Hebrew prophets wore the aba as their attire, according to the Bible. Traditional versions are made of cream-colored wool decorated with striped or embroidered designs in bright colors. An outer garment is worn in the Arab world throughout the Middle East that is voluminous.

There are many types of jellaba in the Arab world. In general, there are wide, long sleeves on the garment, and the skirt with side slits and as a coat or caftan with front placket was used along with it (Figs. 13 and 14).

In addition to outer garments and cloaks, some wear coverings on their heads. Haiks were oblong, striped fabrics with an approximate size of 18 feet by 6 feet (5.5

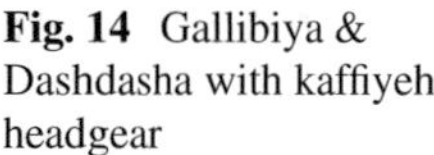

Fig. 14 Gallibiya & Dashdasha with kaffiyeh headgear

by 1.8 m). The material measured about 5.5 by 1.8 m and was worn around the body and head by Arabs at night and during the day. An identical mantle was the burnoose, a hooded garment that was worn day or night to maintain warmth.

Loose, baggy trousers are still quite popular throughout the Middle East, the Balkans, and Anatolia, and are traditionally worn by both sexes. They were drawn tight by cords and measured around 3 yards (2.75 m) at the waist. Depending on the country where they were worn, it was also known as chalvar or shalwar. Ankle ties were used to secure each leg. A thick sash encircled the waist and finished off the chalvar.

This garment was great for field labor since it allowed for freedom of movement and provided protection for the lumbar region of the spine, especially when bending. Men in the military forces have also worn the robe during centuries of conflict. Working clothes are usually made of cotton, although fashionable ladies wear brocade or silk chalvars over linen drawers. Women wore veils or cloaks.

The kaffiyeh has been the traditional headdress of Arab males for centuries. It's still fashionable now, however it's frequently worn with a work suit. Kaffiyehs were traditionally constructed from squares of cotton, linen, wool, or silk folded into triangles and worn on the head with one point on each shoulder and the third falling down the back.

5.2.3 11–15th Century

During the period between the ninth and eleventh centuries, there was a second period of development and wealth. Jeweled embroideries and deep colors, especially purples and reds, were developed into a refined kind of court attire. The royal gown

Fig. 15 Turkish men's traditional

had gold-embroidered panels that wrapped around the body and hung from one arm. The classical style has been replaced by an Eastern-type design. The caftan has evolved into a formal outfit, among other things. The open front of this coat-like garment was fashioned to match the back. Although both sexes wear caftans with trousers, they are not as full as the Middle Eastern chalvars. Instead, they're cut much more gracefully and tucked neatly into boot tops or worn overshoes.

Turkey

As well as the jacket known as şalvar, a cut that covers the lower half of the body with loose pants, a sash called kuşak accompanied the trousers. The caftan was a long robe, decorated with fur borders, kalpak, and sarak adorned the head. Caftans with fur lining and embroidery were used by the administrators and the wealthy, while the middle class wore cübbes and the poor wore cepkens or vests with a collar.

Women usually wear Salwar and a mid-calf or ankle length chemise, zibin a fitted jacket along with that waist belt was also used. In Turkey Women's clothing was Salwar and Chemise, which was an ankle or calf-length dress, along with it zibin was used. It was a small fitted overcoat. Caftan was a robe used during formal occasions. Caftan and zibin both were fastened with buttons. These garments were available in bright colors and patterns. Ferace was a long dark robe tied at throat with buttons, which was used to cover the women's clothing completely during outings. As well as for covering her hair and face, females wore a pair of veils (Figs. 15 and 16).

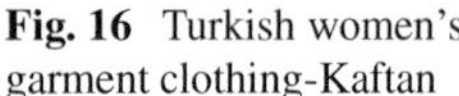

Fig. 16 Turkish women's garment clothing-Kaftan

Medieval Europe

During medieval times, clothing styles were changed and most of them looked uniform, both genders wore an identical wardrobe, though crudely and loosely cut clothes were sewn. Shirts or chemises were worn as underwear as they were a sort of loose-fitting drawers. The attire was mostly made of natural colored linen.

Women's shirts were longer than men's and men's shirts were shorter. The neck was round, the front was slit for ease of donning, and the drawstring fastened the braies at the waist. Usually, one or more tunics were worn over this. Both genders wore tunic as their regular clothing, but the height varies for each gender, where men's tunic falls till the knee or ankle length and women's tunic falls till the floor length. A loose-fitting asymmetrical round-neck tunic with long sleeves was girded at the waist. Tunics made of colored linen or wool have embroidered bands on the hems, wrists, and collars. Two vertical sections of hose were sewed together to cover the man's legs. Garters or bands were used to keep them in place.

The dress of the thirteenth century was characterized by its simplicity. The garments lacked decorations, and the belts were not worn. The tunic was normally worn over a sleeveless surcoat, this was derived from the tabard, a dress style was inspired by the soldiers, where used armor to prevent the sun glare off the metal because during the war period there were chances the enemies may follow because of the metal glare. Women and men often wore surcoats with slits, and a belt was used to attach purse, which helped to safeguard from theft.

The costume changed dramatically around 1350 and outfits became more tailored and shaped to show off the human form. Clothing could be tailored more effectively. Italy and the East were now bringing more and better fabrics to the West. Renaissance culture was perhaps one of the most significant causes of sartorial change. The Renaissance celebrated both the dignity and the importance of humans and revived classical concepts, and by beautifying and displaying the human figure, this was conveyed in the costumes.

In the medieval era, men wore a fitted tunic, which was considered a high-fashion item of clothing. The tunic had four panels, i.e., seams at the sides and back, and it was fastened at the front with buttons. By 1340–45, a lengthy tunic was used which was secured with leather belts, that encircled the hips a few inches above the hem and was embellished with metal and jeweled brooches. Sleeves reached elbow length. From elbow to wrist, long sleeves of the under tunic, of similar cut, were buttoned to fit closely.

Women's dresses also took on new forms. They featured low necklines and straight cuts at the shoulder. A heavy belt fitted around the hips, and the bodice, which reached to the hips, was similar to men's tunic, short skirt with a high waist and long, gathered skirt below the hips. The sleeves were similar to those of men. The top of the gown could also be topped by a sideless surcoat. The armhole and plastron (the front of the gown) were often covered with fur, whereas the gown above had sleeves.

The decorative arts at this time evolved into several new forms. In the first, all clothing, including hose, were of a single color down one side and another across the other, which helped to define the shape of the figure. Changing the ground color and the design color produced counterchange designs that were heraldic, floral, or geometric in motif. There were a variety of shapes were cut in the edge of garments.

These trends continued to develop during the fifteenth century. Hose for men became even more fitted. Suits were often shortened to just below the waist. The material was richer and more elaborately patterned. Over the tunic, long gowns were introduced for older men whose figures were less suitable for display. The dressing gowns were initially full and long (in the fourteenth century), but with vertical pleats in the back and front, they gradually became more tailored and formal. For warmth and appearance, all garments were interlined or edge decorated with fur.

Women wore a wide range of headdresses. The hair coiled over the ears; it was long and plaited. Metal mesh nets encasing the coils were called cauls and were worn with veils. It was fashionable to wear turbans, which were of Byzantine style and introduced to Italy in the fifteenth century. The popularity of witches' hats and short fez-like headdresses resembled that of dunce caps and steeple headdresses. The clothing was designed with beautiful fabrics along with wavy, flowy veils (Fig. 17).

Fig. 17 Surcoat with lengthy armscye and side slit and women's clothing

5.2.4 16–20th Century

Europe

Sixteenth century was the starting point for the further changes in Europe. In France, Flanders, England, and Spain, the Renaissance was gradually laying claim to concepts that had been rejected by medieval societies. The middle class expanded, and people expected a higher standard of living. Europeans began to look abroad as well. Sailing ships sailed to both east and west from Portugal, Spain, and Italy. They acquired richness, precious metals, and new materials on their journeys. All of these things played a part in their clothing.

In fashion, the pacesetters dominated wealth. The style originated from Italy until about 1510. In the following decades, the Germans and the Flemish set the pattern, but from the middle of the century onward, Spain dominated. Early style development reflected the expansion of Italian styles from the late fifteenth century. Shirts worn by youngsters were white and frilled with embroidery at the neck and wrist. A striped tunic and hose were separated from masculine and feminine limbs.

Long dresses, open the center front and showing off the contrasting lining, covered the tunic and hose worn by older men. There were long, flowing hairstyles for men. They wore black velvet hats set at an angle, decorated with brooches and plumes, and decorated with feathers. Dresses for ladies were square-necked, with low-cut bottoms that revealed chemises beneath.

It was common to display skirts under gowns by holding or pinning them up. In the period between 1520 and 1545, padded puffs with decorative slashes shaped the fashionable shape. According to tradition, this idea was taken at the clothing of Swiss culture. People slashed each garment to show the contrast between the color of the undergarment and the one above it.

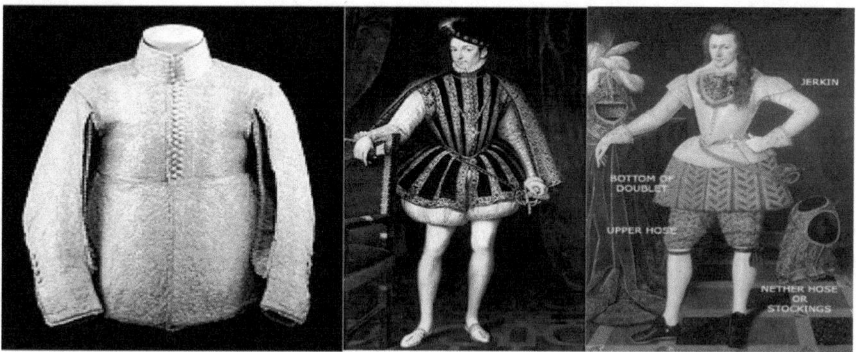

Fig. 18 Doublet with knee length gored skirt

Compared to the humanist fashion of the Renaissance, which displayed figures and was elegant, the new styles were influenced by Northern European Reformation, evident in darker colors, heavier fabrics, and bulky garments padded to disguise figures. Men's tunics, called doublets, wore a knee-length, gored skirt with a protruding codpiece, which was open in front. Approximately half of the gown was covered in a velvet robe with fur collars and padded sleeves. All garments, shoes and boots had wide toes and were slashed in decorative ways. A fashionable hairstyle, a beard, and a cap worn at an angle were short hairstyles, small beards, and flat velvet caps (Fig. 18).

Women's figures were manipulated artificially with a fit underbody with steel or whalebone strips inserted in the side seams to show the waist and torso as slender. Corsets were resulting from this. A wicker hoop was inset at intervals with canvas hoops to give the skirt its cone or inverted-V shape. By 1500, this style was fashionable in Spain, where it originated in the previous century. Nevertheless, the Spanish skirt, known as a verdugado, was shaped like a bell. The cone-shaped hoops were introduced to France about 1530. The queen popularized these hoops and called them ver-tu-gades. So, Farthingales fashion spread and appeared soon after in England (Fig. 19).

During the nineteenth century, the first corset was evolved and female clothing grew as elaborated with boned styles. Farthingales dresses changed as padded sausages were added by 1580, known as bum rolls or barrels, and they were knotted at the waist as inner layer to the skirt. As time went on, the French created the wheel farthingale, shaped like a drum with radiating spokes. A very décolleté neckline was created, almost exposing the breasts. A stomacher is a rigid panel, shaped like a V or a U, and it was heavily decorated with ornaments and embroidery.

During the 1620s, Spanish influence was seen in the Netherlands' clothing. While fine fabrics were still used to make the garments worn by the rich and they were used wool, velvet, and silk in their clothing. The most distinguishing item was lace used widely in bands, collars, and cuffs (Fig. 20).

Fig. 19 Farthingale dress with decorative v-shaped stomacher

Fig. 20 Inner skirt for farthingale garment

The early years of English colonialism were difficult for many colonists. People cultivated flax and cotton and raised sheep for wool to make their clothes. In England, the average person wore plainer clothes than those in the United States. Most colonists wore old-fashioned clothes and kept their finest clothes for Sundays and holidays, which lasted for years. It was conventional for men to wear breeches that were full at the waist, doublets, jerkins, and loose-over garments that were popular in Europe during the late sixteenth century.

A short gown with a full skirt was worn by women over a hand-spun petticoat, which was excellent in quality, a long-lasting garment set. Elegant gowns had longer skirts and were composed of finer materials. Virago was a popular sleeve style among American children and women, with wide sleeves at the elbows and shoulders and tight ribbon drawstrings.

Fashionable clothing was popular among Americans in 1700, and there were no longer any obvious distinctions between the other part of America. The rich and well-to-do Americans dressed according to the prevailing fashions, and there was a difference between city rural areas. In contrast, the former was able to afford expensive fabrics and follow fashion trends, but the latter still wore homespun and woven clothes. The availability of fashion dolls and costume plates made it possible for Americans to stay in touch with the latest styles. Fashion styles were updated from eighteenth-century fashion. Fashion styles were styled and fitted properly. Wearing trousers with shirts became the style of pioneer women and became similar to the men's style. Garments with several layers were reduced to only a few layers. Fashions that fit the body were introduced.

5.2.5 Twenty-first Century

Twenty-first century is considered as modern century. In a few countries, their traditional clothing styles are still followed from the olden centuries. Their cultural clothing styles are preached from their elders to the younger generation. Each generation very carefully passes their traditional clothing styles to the next one. So the cultural garments are still worn on special celebrations. China, Japan, Korea, Thailand, Sri Lanka, India are the countries that still carefully safeguard their cultural clothing.

Japan

The kimono is the traditional costume of Japan. This garment is generally made with silk. The kimono is a loose-fitted full-length garment, has a large, free-size sleeve. Obi is a wide belt used to tie at the waist to hold the garment in position.

The kimono is Japan's traditional attire. Kimonos have long sleeves that span from the shoulders to the heels and are usually made of silk. They are fastened by an obi, which is a wide belt.

During the spring and summer, a special kind of kimono is used by them, which is termed as Yukata. It is an informal kimono, generally available at a cheaper cost than the traditional kimono. This Yukata is a garment for summer, so it is made with lightweight cotton fabric and is available in brighter colors than the traditional one. Yukata is worn during festival and cherry blossom viewing ceremonies. Hakama is a long pleated skirt and it is worn along with the kimono as a formal wear. Kimono with Hakama is originally designed for men, but nowadays it is worn by both sexes.

Obi is a belt-like accessory that wraps around the outer kimono and helps to keep all the layers together but does not close it. It is typically long and rectangular and can be decorated in several different ways. Obi is made from many different fabrics. The fabric used for modern obi is typically crisp, if not stiff, and can be relatively thick and unpliable.

Wearing traditional Japanese clothing creates a lot of fun, and renting kimonos while sightseeing has become increasingly popular in recent years. Japan calls traditional clothing wafuku, meaning "Japanese clothing," to distinguish them from others. Nevertheless, there are many different kinds of wafuku, so here we will discuss the main types of clothing that you are likely to wear if you live in Japan.

Fig. 21 Children's Kimono, Spinster's Furisode, Women's kimono, Bridal attire

There is information in this section about clothes for men and women, formal wear and casual wear, and different occasions and seasons.

Below are a few basic kimono types:

Furisodes are long-sleeved kimonos worn by young female singles at formal events that are formal like weddings or puberty ceremonies. Silk Furisode is a very colorful fabric with bright designs. Homongi is a semi-formal kimono fitting for both single and married women. It would be worn for a tea ceremony or a wedding.

The komon kimono is a more casual version for women that features a finely detailed pattern that is repeated numerous times. These shoes come in a variety of shapes and colors, and they're perfect for wearing about to town. When renting a kimono in Japan for touring, the kimono is most often a komon.

Men's kimonos have less diversity than women's kimonos. Both men and women wear kimonos with hakama skirts and haori jackets for formal occasions. Kinagashi refers to kimonos worn without a hakama by men. The level of formality for women is determined by the grade of material used, for instance, fine silk for formal occasions, and lower-grade tsumugi silk for less formal events.

Nagajuban is a type of undergarment worn under a kimono in Japanese culture. The top and shorts are made from linen or cotton and are similar to western pajamas. In addition to being worn around the home, they are also commonly worn on summer festivals by men and children. Summer festivals are also a place to wear bright colored jinbei.

The width of Kimono cloth is typically 14 inches. In the United States, 45 and 60 inches are comparable measurements. Bolts of kimono fabric include around 12.5 yards of cloth, which is enough to make an adult-sized kimono. There are four elements that all kimonos have in common. Standard fabric widths are utilized in geometric forms, requiring little trimming. The front is asymmetrical and exposed. The front aperture is fastened with an elastic band. The kosode and kimono do not have any closures (Figs. 21 and 22).

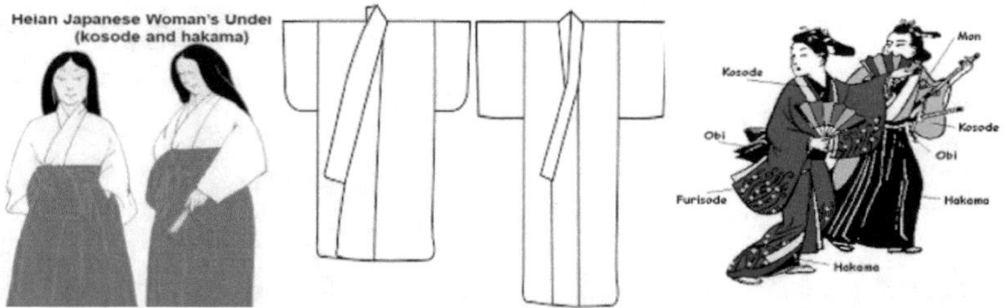

For the wedding, the groom wears a black kimono with the family crest made of habutae silk, a hakama (a fleeted skirt), and a half-length black coat called haori.

Funeral—The kimono's basic cut is similar for men and women. The bottom outside corner of the sleeve differs in shape, which is the fundamental gender difference in the kimono. Adult women's kimono sleeves are somewhat rounded, whereas men's are square cut. Children's sleeves are the most rounded, followed by single women's sleeves.

Thailand

Traditional Thai garment patterns include Thai silk and ankle-length wrap and pasin. Royal type national clothing comes in eight main variations. Thai Borom Bimarn, Thai Chakri, Thai Dusit, Thai Chakraphat, Thai Sivalai, Thai Reun Ton, Thai Chitralada, and Thai Amarin are the traditional styles of Thai clothing.

Fig. 22 Thailand traditional dress

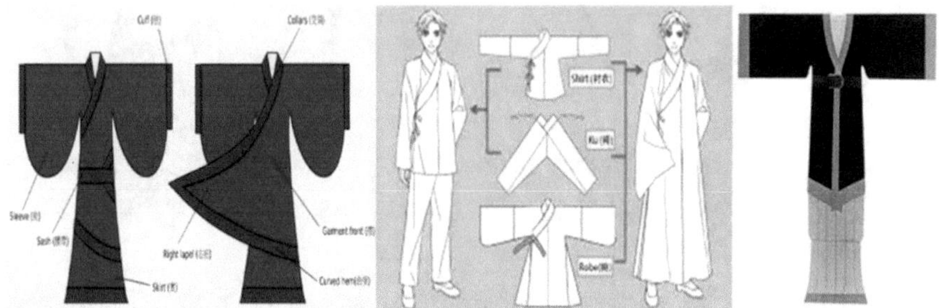

Fig. 23 Chang & Bixi, White cotton inner wears, Kuzhe: a short coat with trousers

China

Traditional clothing is only worn at certain festivals, rites, or religious occasions. Many of the country's ethnic minorities continue to dress in their traditional costume daily, and they played an important role in traditional Chinese clothing. To make an item of ancient Chinese clothing "semi-formal," add the following elements: Chang: a pleated skirt with a long front cloth panel attached to the waist belt.

Bixi: a pleated skirt with a long front fabric panel linked to the waist belt. Zhaoshan: a long coat with a front open, a Guan or other ceremonial hat; this outfit is appropriate for welcoming guests, attending meetings, and taking part in other important cultural activities in general. Because these are often expensive garments made of silks and damasks, they are generally worn by aristocracy and upper-class people.

Coat sleeves are usually longer than shenyi to give the impression of larger volume. Other Ethnic Minority Clothing: There are 54 ethnic minorities in the country, each having its own set of costumes. Unlike traditional Han Chinese costumes, these costumes are still fashionable today. The two possibilities are a long gown and a short coat with pants or a skirt. Long gowns with hats and boots are common; some choose short coats and, in general, wear shoes and wrap their heads in a cloth (Figs. 23 and 24).

In the Ming Dynasty, it was a formal dress worn by intellectuals and students sitting at the imperial examination. It has a round neck with a button and is wide-sleeved with black edges. It must be worn with a crossed-collar undergarment underneath. Side slits may or may not be present (with side panels to conceal the undergarment). Since the Tang Dynasty, it has been worn (Fig. 25).

Sri Lanka

Sri Lanka Longyi—checks, plaids, or stripes in any color that run the length of the fabric. Two meter-long cotton garment with one end twisted into a half-knot and tucked at the waist. Hetta is a fitted, short-sleeved shirt with a waist length. It may

Fig. 24 Old & current clothing style of chinese woman

Fig. 25 Recent days Chinese traditional wear

or may not have buttons on the front, but it invariably has a closed back, unlike the choli. Cheeththa is a long ankle-length skirt or a skirt that wraps around the body. A sort of sari is the osari. Redda is a wraparound skirt with a cap tucked in the waist. A two-piece dress with long ruffles at the shoulders and waist is known as a lamsari (Fig. 26).

India

India is famous for its traditional clolorful clothing. Each state in India has its own style of traditional clothing. Sari and choli are our common garments. Sari is a rectangle piece of cloth commonly worn in India by women. The length of the sari varies from 5.5 yards to 9 yards, its length and draping style various from state to state. Half sari, front pleat and back pleat style, madisar style, and pallu at right arm style are the various styles of sari draping. Choli and blouse with an othni are traditional costumes of women in North India. Men's costumes are dhoti with a shirt or salwar with kurtha. Cotton clothing is found everywhere due to climatic conditions. Wearing of turban on the head is seen in all states especially labors who work under direct sun. Silk is considered a holistic garment for a special occasion.

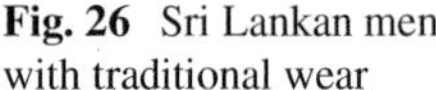

Fig. 26 Sri Lankan men with traditional wear

In Tamil Nadu, the sari is a very important piece of traditional women's apparel. Kancheepuram saris are well-known for their excellent elegance. They differ from North Indian saris in terms of color, texture, and style. The sari's length is usually between five and six yards (Figs. 27 and 28).

Women wear blouses to cover their upper bodies. The blouse covers till mid-torso. It may be stitched however the wearer desires it to be with different types of color work, lengths of the sleeve, and even the length of the blouse may vary.

Both men and women wear the Lungi in Kerala, which is also known as Kaili or Kalli Mundu. It is considered working or informal clothing for laborers. Lungis are worn by the majority of men in Kerala as house or sleepwear. The state's traditional attire is known as "Mundu," and it is worn on the lower half of the body, from the waist to the foot. It's white, and both men and women wear it. It has the appearance of a long skirt or a dhoti. Gender and age affect the upper garment. Mundu is a white cloth wrapped around the waist as a lower garment. It has a Kara border which can be in any color, but it is usually golden. Kara gives the Mundu a flair by showing it on the person's left or right side.

Fig. 27 Tamilnadu's saridraping style

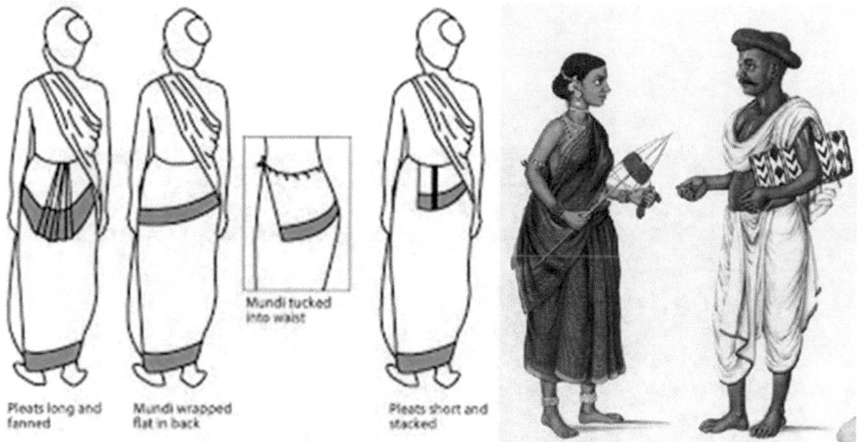

Fig. 28 Traditional sari draping method

The upper garment is called "Melmundu," and it is worn on the shoulders like a towel. Many males these days wear white shirts as well. Women's traditional clothing is known as "Mundum-Neriyathum." This is made up of two Mundus that are very similar. One of them is worn around the hips and reaches the ankles on the lower half of the torso (Figs. 29 and 30).

Women in Maharastra wear nine-yard-long saris. The sari is tucked in the middle, giving it a dhoti-like appearance. The remainder of the sari is draped over the woman's upper torso. Lugade is the name given to this sort of sari. The sari was worn in a variety of ways by women (Fig. 31).

Fig. 29 Kerala Men's wear Dhoti & Mundu

Fig. 30 Maharastra Women's clothing

Knife pleated skirts, box pleated skirts, inverted box pleated skirts, six-panel petticoats, elastic plain skirts, Patiala pants, wraparound skirts, and kimono nighties are all common. These are zero-waste cut clothing that has been around for a long

Fig. 31 Various states Saree draping style in India

Fig. 32 Varieties of zero-waste clothing

time. This clothing has been worn in most countries throughout history, both before and after the Common Era (Fig. 32).

6 Impact on Sustainable Fashion

Zero-waste fashion refers to clothing that generates very little or no textile waste during production. It is within the umbrella of the greater sustainable fashion movement. It can be classified into two groups. Waste is eliminated before it reaches the consumer hand. Post-consumer zero-waste fashion uses post-consumer garments, such as second hand clothing, to create apparel that reduces waste at the end of a garment's product life cycle. The concept of zero-waste fashion is not new [5], with the kimono, sari, chiton, and a range of other traditional folk costumes serving as early instances of zero-waste or near-zero-waste clothes.

This category encompasses two broad techniques, both of which occur during the garment's initial production. Working within the cloth width, the designer creates a garment utilizing the pattern cutting method in zero-waste fashion design. [2] This approach has a direct impact on the finished garment's design because pattern cutting is a vital design stage. Although drawing can be a great exploratory tool, designing a zero-waste garment only through sketching is tough. Zero-waste production, which includes zero-waste design, is a comprehensive approach to eliminating textile waste without requiring changes to clothing patterns. The three "R"s of the trash hierarchy are Reduce, Reuse, and Recycle, in order to reduce environmental impact. Textile waste is removed before it reaches the consumer in zero-waste fashion design, but it does not necessarily address waste generated during the garment's useful life and disposal.

7 Impact of Slow Fashion

Slow fashion is an antithesis to fast fashion and part of the "slow movement," which advocates for manufacturing that respects people, the environment, and animals.

Slow fashion, as opposed to fast fashion, involves local artisans and the use of ecologically friendly materials, intending to protect crafts and the environment while simultaneously offering value to both consumers and producers. According to slow fashion, "Identify sustainable fashion alternatives, based on the repositioning of design, production, consumption, use, and reuse processes, that are emerging alongside the global fashion system, and are providing a potential challenge to it."

It is an alternative to fast fashion in the sense that it encourages a more ethical and sustainable way of living and purchasing. [3] "All facets of the 'sustainable,' 'eco,' 'green,' and 'ethical' fashion movements are included." [4] The Slow Consumption Movement is another business approach that focuses on both slowing down consumerism and conserving the environment and ethics. For a long time, it was linked to a distaste for quick fashion.

Slow fashion, as opposed to fast fashion, emphasizes high-quality production to lengthen the life of the garment or material. The purpose of slow fashion is to create a garment with a cultural and emotional connection; buyers will keep an item of apparel for more than one season if they are emotionally or culturally attached to it. [8] A tax is being devised to make it more difficult for fashion companies to buy or produce materials that are not made from recycled, organic, or reused materials. [9] The industry's carbon footprint will be decreased by repurposing previously made resources [9].

There is also a strong drive for businesses to be more transparent. Many sustainable fashion firms make their clothing and design manufacturing methods public, allowing customers to make more educated purchasing decisions. [10] Following the slow movement, there is a trend toward more informed purchasing, as well as new clients for businesses.

8 Conclusion

Slow fashion is gaining steam, fueled by increased environmental concerns, as part of a greater goal of sustainable fashion and, as a result, cleaner earth. Slow fashion is a prominent counter-trend to fast fashion. Slow fashion, she claims, is more concerned with quality than with time. Other early adopters of the slow fashion movement point out that it promotes slower production, integrates sustainability and ethics, and eventually encourages buyers to spend on well-made, long-lasting clothing. Slow fashion and zero-waste apparel aren't exactly new concepts. It had been widely embraced only a century ago. This culture has lost its luster over the last 100 years, i.e., the twentieth century. Now, due to increased environmental awareness, it has been loudly proclaimed once more. Slow fashion adaptations of traditional garments will protect the lives of all living things on the planet.

References

1. Brian P (2005) Japanese childhood, modern childhood: The Nation-State, the school, and 19th-century globalization. *Journal of Social History* 38(4):965–985 « in JSTOR
2. Kathleen SU (1999) Passages to modernity: Motherhood, childhood, and social reform in Early 20th century Japan
3. Mark J (2010) Children as treasures: Childhood and the middle class in Early 20th century Japan
4. David S.N, Arthur FW (eds) (1959) Confucianism in action, p 302
5. Minute on Education (1835) by Thomas Babington Macaulay. www.columbia.edu. Retrieved 3 May 2016
6. Latika C (2010, April–June) Land revenues, schools and literacy: A historical examination of public and private funding of education. Indian Economic & Social History Review 47(2):179–204
7. Bhutani VC (1973) Curzon'S educational reform in India. Journal of Indian History 51(151):65–92
8. Leaving school before the age of 16. Education.govt.nz. 24 May 2017. Retrieved 1 December 2017.
9. McCreadie M The evolution of education in Australia. IFHAA Australian Schools
10. U.S. DOE (1959), 1, cited in Paglayan 2021
11. Jump up to:[a] [b] Wolf-Gazo, John Dewey in Turkey: An Educational Mission, 15–42.
12. Republic of Turkey Ministry of National Education. Atatürk's views on education. T.C. Government. Retrieved 20 November 2007.
13. Wolf-Gazo E (1996) John Dewey in Turkey: An educational mission. J Am Stud Turk 3:15–42
14. Reisman A (2007) German Jewish Intellectuals' Diaspora in Turkey: 1933–55. Historian 69(3):450–478
15. Arnold R, Jewish Refugees from Nazism, Albert Einstein, and the Modernization of Higher Education in Turkey (19).

Advancements in Recycling of Polyethylene Terephthalate Wastes: A Sustainable Solution to Achieve a Circular Economy

G. Jeya, T. G. Sunitha, V. Sivasankar, and V. Sivamurugan

Abstract Using plastics has become an essential part of human life today and plays a pivotal role in food packaging to fabrics. Among plastics, polyethylene terephthalate (PET), also known as "polyester resin" or PES fibers, has made it possible for lower-income communities all over the world to develop cost-effective fabrics. Since then, it increased tremendously the usage of polyester resin in either pure form or blended with other fiber form and over 60% clothing produced using polyester resin. The life cycle assessment of PET fibers revealed that in order to reduce the environmental impact, effective recycling methods must be implemented. Physical and chemical methods, which reduce the use of fresh raw materials, could majorly recycle the post-consumer waste of PES fabrics or fibers. Most recently, the development of the circular economy has created an ample opportunity for "recycled PES" in textile fibers. The chapter intended to provide information on recent developments in the physical and chemical recycling of PES fibers or any PET waste such as beverage bottles and the conversion of them into PES fibers. In addition, we have discussed the degradation of PES using biological treatment methods and biodegradable PES fibers in the chapter.

Keywords Aminolysis · Biological treatment · Chemical recycling · Circular economy · Depolymerization · Glycolysis · PET · Textile wastes

1 Introduction

One of the largest consumers of natural and synthetic fibers is the textile and apparel industries. PET, also known as PES fibers in the textile and apparel industries, is a polyethylene terephthalate (PET). The global production of synthetic and natural textiles jointly, as estimated in 2017, is up to 94.7 MT. According to the JCFA 2017 report, PES alone contributes to 25.4 MT and plays a major role in synthetic fiber market [47, 56]. Because of the large volume of production of PES synthetic fibers,

G. Jeya · T. G. Sunitha · V. Sivasankar · V. Sivamurugan (✉)
Research Department of Chemistry, Pachaiyappa's College, Chennai, India

S. S. Muthu (ed.), *Sustainable Approaches in Textiles and Fashion*, Sustainable Textiles: Production, Processing, Manufacturing & Chemistry,
https://doi.org/10.1007/978-981-19-0530-8_6

the management of solid waste generated after consumption is a major problem for disposal. The recovery of material from post-consumer PES textile waste through physical, chemical, and biological methods would contribute to the conservation of raw materials of petrochemical origin (Fig. 1). Unfortunately, the material recovery from waste PES has not closely followed by many industries. Unlike the PET bottles have collected by local, municipal, and metropolitan authorities through strict environmental policies, post-consumer PES waste has not focused. In addition, there are PET recycling companies that contribute to material recovery, preferably through physical recycling, while chemical and biological recycling process is adopted by few industries. However, no roadmap is available for the collection of end-user PES textile waste for subsequent reusing or recycling.

In conventional method, most of the textile wastes are collected along with other common plastics incinerated or disposed as landfills. Due to the traditional technique, a large volume of textile waste discarded of as landfills has become a source of microplastic pollution in soil due to a higher surface area and moisture absorption potential. However, the vision of achieving circular economy and reducing carbon and water footprint through sustainable process is emerging among textile and apparel industries to meet legislative and customer requirements [97].

Traditional disposal methods, such as landfilling or incineration, have negative environmental consequences. The majority of polyester fibers are obtained from petroleum-based chemicals. Recycling of PES textile waste drastically reduces the consumption of conventional petroleum-based products. Compared to mechanical or

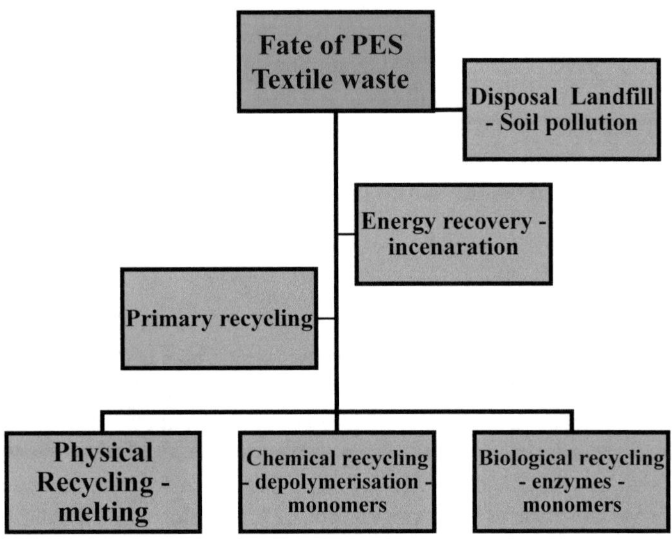

Fig. 1 PES Textile wastes recycling methods

physical recycling, chemical recycling processes deliver pure raw materials (derivatives of terephthalic acid monomer) for the synthesis of polyester fiber [71]. Alternatively, PET could be depolymerized by enzymes, which is a more environmentally benign process [2154, 114].

It identified one source of microplastic pollution in the air as textile fabrics [35]. The improper disposal of PES textile wastes could be a potential hazard for causing microplastic pollution [25, 26]. Recycling PES post-consumer textiles contributes not only to the circular economy but also to reduce microplastic pollution [77]. Most common benefits of using recycled PES, which significantly reduces CO_2 emissions, consumption of petroleum-based feedstocks, and solid wastes. The majority of the apparel industry tend to favor textiles made from recycled PES. Major giants, namely Adidas, Nike, Ikea, Coca-Cola, Walmart, and more other industries, have taken sustainable initiatives to reduce carbon footprint as their corporate social responsibility and long-term business vision [73]. Relatively, the recycled PES obtained from pre-consumer PES fibers is more economical compared to post-consumer PET fibers because of the extensive cleaning and removal of dyes/impurities required. In chapter, we elaborated generation, chemical and biological treatment PES textile wastes. The chapter briefly discussed contribution of textile wastes to circular economy and methods of achieving circular economy through physical, chemical, and biological treatment. Sources of textile wastes and various methods available for recycling of textile wastes are highlighted in the current chapter.

1.1 Current Scenario of Textile Wastes

Recycling of textile waste materials such as PES at any stages of life cycle analyses would reduce use of traditional petrochemical raw materials and reduce its disposal [93]. When compared to physical or mechanical recycling, the chemical recycling process is more beneficial. In the chemical recycling process, post-consumer textile waste depolymerized to monomers and charcoal-based adsorbents could remove impurities (Fig. 2). We could use the pure depolymerized monomer for the synthesis of PES fiber production. There are many industries that produce recycled PES yarn under different commercial names. The major players in the textile industries preferred to use recycled PES collected from either post-consumer PET bottles or pre-consumer PET waste from textile industries. We could also recycle the textile wastes using biological pathways [75].

Major brands such as Adidas, Nike, H&M, Puma prefer fabrics obtained from recycled PES as their company's sustainable policy and social responsibility. ECOTEC® is the brand name of recycled PES yarn produced by Marchi & Fildi, Italy. The recycled PES obtained from pre-consumer PES waste yarns was generated during the production of the manufacturing process. PHOENIX® yarn is made from recycled post-consumer PET waste.

As per the prediction by Greenpeace forecasts, the amount of PES clothing will be doubled by 2030. According to the Intergovernmental Panel on Climate Change,

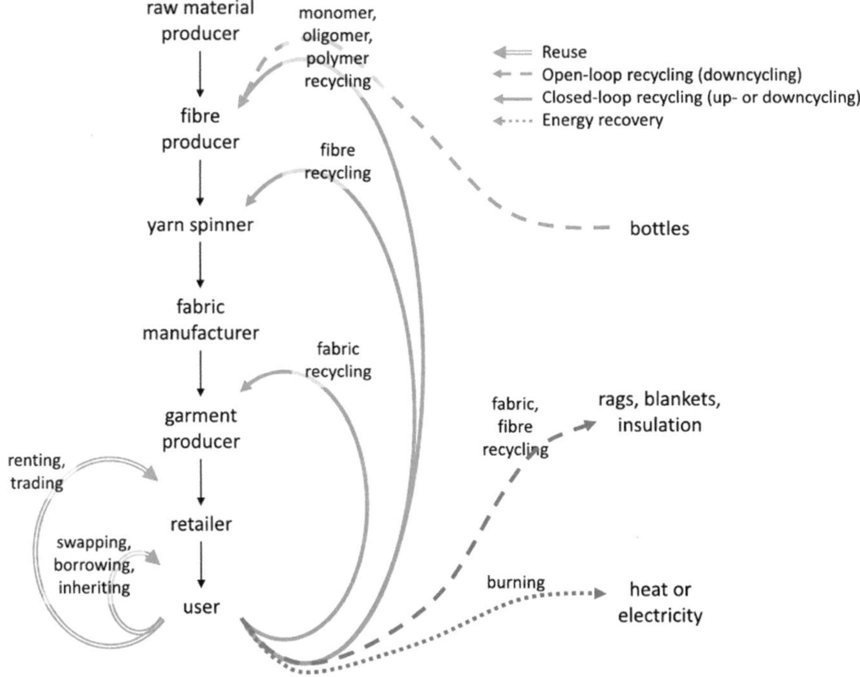

Fig. 2 A classification of textile reuse and recycling routes (*Source* [86]. This article is distributed under Creative Commons CC-BY license)

traditional PES production uses crude oil-based raw materials, which may have a negative impact on GHG emissions and global warming (IPCC). There are PES textile and apparel giants such as Adidas, H&M, GAP, and IKEA which used recycled PES for clothing. They see to increase up to 20–25% recycled PES in their garment production. According to the report, the textile and apparel giants showed great interest in implementing recycled PES in their textiles and exceeded their target. However, most companies mainly use mechanically or physically recycled PES obtained from post-consumer PET bottles, rather than PES fibers. Some note that PES fibers derived from physical recycling of five PET bottles could make a T-shirt. However, physical or mechanical recycling of PES fibers may lose their quality during subsequent recycling because of high temperature. The addition of recycled PES with virgin PET has some advantages, such as reducing the plastic production, which usually disposed as either landfills or in ocean, and raw materials consumption. WRAP is an organization, which provides sustainable solution to the industries for improving resource efficiency through collection and reprocessing of food, drink, clothing, and textiles. According to the WRAP report, the use of recycled PES reduces up to 32% of CO_2 emission when compared to the use of virgin PES.

Nevertheless, there are some constraints to using recycled PES, including the use of PES blends rather than pure PES in the textile industry, the removal of dyes and other ingredients, and the contribution to microplastic pollution disposal problems. Despite having difficulties, enzyme-based techniques have developed for the selective separation of cotton and wool from the PES blends [64, 76]. Haren textiles in Mumbai, India, produce recycled PES fibers from recycling PET bottles. Libolon from Taiwan produces recycled PES yarn under the name of RePET, which is obtained mainly from post-consumer PET bottles. Monks International of Belgium produces recycled PES yarns using mechanical recycling of waste PET bottles. Interestingly, Unifi from USA produces REPREVE® which is recycled PES obtained from PET wastes disposed into landfills or in ocean.

According to OECO textile reports, almost 65% of textile segment engaged by synthetic fibers and within that segment PES covered up to 60%. Using recycled PES in textile industries is one of the sustainable initiatives since it consumes less energy for production compared to the production of virgin PES. Using post-consumer PET bottles reduces its disposal as landfills. According to a report, 125 MJ/kg is consumed to make virgin PES fiber, whereas the conversion to make recycled PES fiber consumes only 66 MJ/kg. Majority of recycled PES yarns are obtained from physical or mechanical recycling. Thus, the quality of PES gradually decreased with the subsequent recycling process. In terms of moisture transport performance, both virgin PET and recycled PET fabrics were good [43]. For sports clothing, the ability to transport liquids is critical. This property has been evaluated by analyzing dynamic liquid transport, capillary transport drying, and water absorption capacity.

2 PES Textile Waste

Conventional PET creates pollution along its entire value chain during the production, use and end-of-life phases, and also contributes to the unsustainable depletion of nonrenewable resources. PET garment consumption degrades environmental quality, destroys ecosystems, and endangers human health [68]. The textile waste originates from the textile manufacturing process, pre-consumers, post-consumers, and through the commercial and service industries [18]. Textile waste becomes worthless after the production process and are classified into pre-consumer, post-consumer, and industrial textile wastes based on the generation.

2.1 Pre-consumer Textile Waste

Textile scraps after cutting a garment piece, leftover textile samples, selvages, end-of-roll wastes, damaged materials, and partially finished or finished clothing samples from the design and production departments are examples of pre-consumer waste [24]. The recycling of pre-consumer waste is economical as well as viable, since the raw materials used during manufacturing are exactly known. The pre-consumer waste can be minimized by using proper computer-aided design software of the textile manufacturing instruments, which reduces the size of cut-out pieces and reduces the quantity of pre-consumer waste, which is a good example of sustainable manufacturing.

2.2 Post-consumer Textile Waste

Post-consumer textile waste results from the finished products, which are thrown away by the consumer after completing its life cycle. The sorting and collection of post-consumer waste is difficult, and the chance of post-consumer waste ends up in landfill or oceans and pollutes the land, water, and air. The conventional PET is not biodegradable and leads to fragmentation over time, releasing harmful additives and microfibers [62].

Incineration of PET textile waste produces CO_2, CO, benzene, TPA, benzoic acid, acetaldehyde, aliphatic hydrocarbons, and smaller amounts of dioxins and furan. These compounds cause environmental hazards and a threat to human health. The recycling of textile waste to recover polymers or even monomers avoid the need for incineration or landfilling [91]. The properties of recycled PET post-consumer waste decreased because of their lower ceiling temperature and their sensitivity to moisture and alcohols [122].

2.3 Industrial Textile Waste

Industrial textile waste is referred to as "dirty waste" because it is generated from commercial and industrial textile applications, such as waste carpets and curtains. A significant proportion of these end-of-life goods are incinerated or dumped to landfills. These textile wastes are inevitable and open to downcycling, upcycling, and recycling to incorporate the waste into an asset, thus leading to a circular economy [40].

2.4 Downcycling

Downcycling is the process of converting a high-value material into a low-value product that is mostly used for cleaning. This includes reusing old clothes to make non-woven textiles, building insulation, rags, or carpet underlay. The textile and clothing, after extensive use, would mostly have short fibers which are usually downcycled into insulating or filling materials. These products, after their useful life, cannot be used again in any form, and thus they can be used for landfill.

2.5 Upcycling

Old materials from pre–consumer or post–consumer waste, or a combination of the two, are used in the upcycling process. The upcycled product will be more valuable and used for home furnishings, artwork, and, most recently, fabric manufacturing. Because there is no quality loss during the recycling process, these products have a much longer useful life.

2.6 Recycling and Its Challenges

The recycling of polyester waste gives a secondary source of raw material to manufacture similar or dissimilar products, thus avoiding the need for virgin raw materials. Old materials from pre–consumer or post–consumer waste, or a combination of the two, are used in the upcycling process. The upcycled product will be more valuable and used for home furnishings, artwork, and, most recently, fabric manufacturing. Because there is no quality loss during the recycling process, these products have a much longer useful life. Attempts have been made to develop automatic sorting process for fabric wastes using near-infrared reflectance spectroscopy (NIR) [19] and ATR-FTIR method [80]. The carbon black is used in food trays which has high PET content. This carbon black could not be identified by the NIR technique because the black pigments do not reflect in the NIR sorting device,hence, the polymer waste is sent for landfill or incineration. The polyOne Corporation developed a black colorant (OnColor™ IR Sortable Black) by which black polymers are detected by NIR automatic sorting equipment and recycled. However, scalability of such technologies is still challenging because of cost and infrastructure implications. Biosorting techniques employ living organisms or enzymes to consume a portion of the fibers while leaving the remainder behind [17].

The presence of additives, dyes, mordants, and coatings restricts the possibility of fabric recycling [27]. Solvent extraction methods are used to remove dyes from textiles. If the dispersed dyes used in PET fabrics are not removed properly before recycling, it will hinder the usability of the recycled product [23]. Polymer blends are mixtures of different fibers whose compositions and structure vary from product to product. The polyesters can be recovered by dissolution of the blends in suitable solvents after precipitating and recovering the polyester. Some of the fiber blends are difficult to separate from and hinder the recyclability potential of fabric wastes. Although polyester has been the dominant fiber in the textile industry, the technology for recycling polyester fabrics seeks more advancement at workable levels [30].

3 Circular Economy: Sustainable Solution for PES Textile Wastes

A circular economy is restorative and regenerative by design. This means that materials are continuously flowing through a "closed-loop" system rather than being used once and then discarded. It is contrary to a linear economy where the raw materials are recovered from the waste plastics, reused and finally discarded. Plastic recycling is the process of recovering scrap or waste plastic and reprocessing it into new and useful products. Plastic recycling reduces high rates of plastic pollution and conserves resources without depending on or maximizing the use of virgin materials to produce brand new plastic products.

3.1 Open-loop Recycling System

Open-loop recycling refers to a system in which products are reprocessed and the obtained recyclate are used for different applications. The recycled material will not be used again instead disposed at the end of its life because of its modified property which can never replace the virgin material [20].

PET bottle recycling to fiber is an example of open-loop recycling. PET bottles are recycled into PET flakes, which are then converted into fibers. The quality of the recycled PET fibers depends on contaminants, which include dust, coloring agents, acids, and water [6] which were used during the recycling of PET bottles.

The amount of solid waste disposed to the landfill can be reduced by recycling the waste for making new products which could be the feedstock of another. According to life cycle assessments, open-loop recycling of PET bottles has benefited the environment by lowering municipal solid waste and reducing the consumption of excess PET virgin materials.

3.2 Closed-loop Recycling System

The life cycle of a closed-loop system begins with the extraction of raw materials and continues with the design and use of products, collection and recycling, or final disposal, as advocated by the very essence of circular economy. In a closed-loop recycling system the products of comparable quality can be manufactured again using recycled material in place of virgin material. In the context of life cycle assessment, the product of the same quality can be manufactured thanks to the inherent material properties possessed by the recycled material as the virgin material. The circular economy employs the 3R principle (reduce, reuse, recycle) to create a closed loop by minimizing the use of raw materials and achieves economic growth, environmental protection, and social benefits for the plastic problem.

3.3 *Reduce*

Reduce means primary goal to reduce the use of natural resources and energy. The new PET preforms are produced by mixing the re-granulate with the new granulate and melted, then fed into injection molding machines to produce "preforms" for new PET bottles which reduces the use of crude oil. Recycling plastic uses 88% less energy than producing plastic from new raw materials [67]. Using PET bottles as raw material for polyester fabrics provides 75% reduction in greenhouse gas emissions compared to "virgin" petroleum PET (Lorenz, PET Circular Economy) [59].

3.4 *Repair/Reuse*

Repair is a critical component of circular economy (CE) strategies for extending the useful life of products and materials. This refers to the fixing of a specified fault in a textile product to make it fully functional for its originally intended purpose, and thus the product's lifetime is extended. Reuse is the practice of reusing a product or its components after its first life cycle for subsequent life cycles in order to reduce the use of virgin materials in the production of newer products and components. The waste polyester textile fibers have good mechanical properties and therefore can be reused to prepare composites. [116] produced composites from polyester and cotton blend fabrics without the reinforcing materials. By reusing materials, manufacturers are able to get closer to their sustainability goals.

3.5 *Recycle*

Waste plastic can be used to make food-grade PET not only from plastic bottles, but also from waste recovered from oceans and polyester textiles.

There are four plastic waste recycling processes.

Primary recycling:

The primary recycling method is re-extrusion, i.e., the reintroduction of scrap, industrial, or single-polymer plastic edges and parts into the extrusion cycle in order to produce similar material products [3]. It is also referred to as "closed-loop recycling" because it does not require sorting or decontamination. Uncontaminated industrial scrap can be recycled in its entirety or mixed with virgin material and reused for its original application. The primary recycling procedure makes production easier, removes impurities, and increases product stability. Primary recycling is carried out using the quantity of PET recovered from post-consumer bottles toward the production of new bottles which has similar strength and performance characteristics to the original bottle.

Secondary recycling:
Secondary recycling is still a mechanical recycling process. It is also known as "material recycling." The recovered PET wastes, after separation and sorting from contaminants, are being crushed, grounded, and reprocessed to granules [70]. The recycled product does not have the same physical demands as the original product and is often less recyclable. As a result, this recycling method is not used to create products that must meet high quality standards. The heat of fusion causes photooxidation and mechanical stress as a result of the inverse reaction, resulting in a decrease in product quality. The conversion of plastic bottles into polyester fibers is an example of secondary recycling.

Tertiary recycling:
Tertiary recycling also known as feedstock. Chemical recycling has great potential in the circular economy of plastics; it can close the loop by producing starting monomers from the polymers that may be reprocessed to produce high value-added chemicals. Chemical process needs selective and active catalysts that are costly and energy-consuming, and require high temperatures to give a product mixture. However, the product mixture is difficult to separate, limiting its industrial applications [34]. The glycolysis of PET into diols and dimethyl terephthalate, which can be used to re-manufacture PET products or other synthetic chemicals, is an example of tertiary recycling.

Quaternary recycling:
By incinerating PET waste, energy can be recovered from plastic waste. By recovering the chemical energy stored in PET waste, incineration generates thermal energy. It is an unfavorable method because it pollutes the air and poses health risks from toxic gases produced during the incineration of PET waste.

4 Physical or Mechanical Recycling of PES Textile Waste

4.1 *Textile-to-Fiber*

Mechanical methods for re-fiber fabrics have been in use since the industrial reclaimed. Carded yarn spinning and unadulterated yarn are traditional practices for changes wool into reusable or later yarn. The rotor spinning process, thread covering process, and fraction spinning process DREF are among the new processes for growing reinforced fiber, allowing for economic production [38]. Machine recycling machines cut the fabric into tiny fragment and then gradually shedding the fabric until it is in a state of fiber suitable for other process such as spiral yarn or making non-woven textiles (Fig. 3). A regular procedure is the following.

- Fabrics are sorted, the metal component removed, and then the fabrics are sealed.

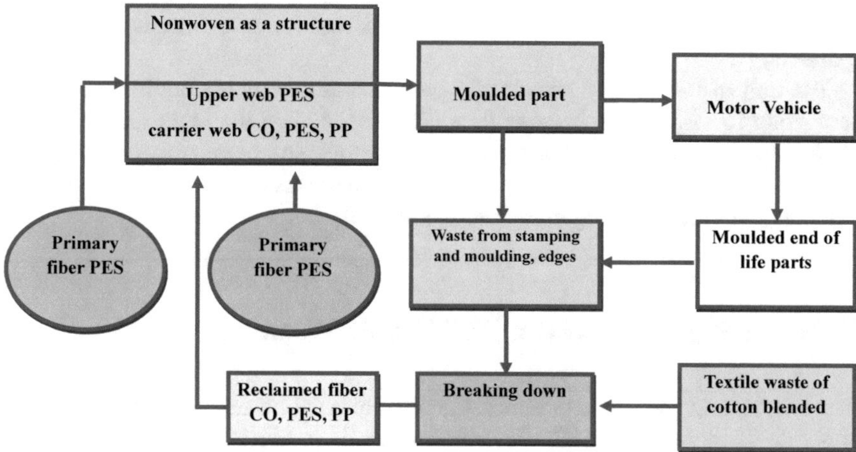

Fig. 3 Production of reclaimed fibers and their reuse in the motor vehicle (*Credit* [38], Permission granted)

- The fabrics should be cut into small pieces by a rotary blade, for example, 1 × 8 cm stripping.
- The fibers are separated by a process called pulling, picking up, or tearing as the fabrics gradually rolled on a small sharp surface to break, they also removed the fibers [58].

The method using to recycling fabrics made from natural fibers etc. cotton and wool, synthetic fibers along with PES, nylon and blended fibers.Acrylic garments are sorted and cut, mechanical shredded into the fiber, and then again woven in the blanket to acrylic yarn [44]. Polyester recovered from mechanical rotation is often used in low-value application because owing to forfeiture of physical properties, degradation and contamination during utility cycles and processing [74].

4.2 *Bottle-to-Fiber*

Bottle waste fibers by open-loop recycling method in which PET is recycled into PET flakes, which are spun into fiber and woven into knitting or textiles. Textile waste, in contrast to the method described above, enters the second product life cycle, and this approach to OLR involves the use of waste from other product cycles in textile and clothing production [72]. Since polymer is non-biodegradable, recycling is an appropriate use of the resources. Bottled had recycled in the 1970s. Regardless, the quality of recycled polyethylene terephthalate is determined by whether the recycled bottles contain contaminants such as dust, color, acids, and water [6]. The mechanical circulation of PET waste to fiber is accomplished as follows: the bottles are collected, cleaned, and sorted by color. Labels are being removed. The bottles are processed into PET flakes, the polymer flake is liquefied, and the new fiber is excreted via spinneret.

The trait of R-PET in mechanically processes is commonly not higher because comparing to virgin PET because the purity depends on the impurities inside the bottles [102]. Like the mechanical process, polyethylene terephthalate could be recycled using a chemical process. Nonetheless, bottled waste, as it may be chemically modified, will return to oligomer or monomer [90].

5 Recycling and Reuse of PES Textile Waste

Based on natural and synthetic sources textile products are categorized. Engineering of textile fibers is a technical term used for textile products. Natural filaments are sourced from plants and animals [29, 108]. Cotton, kenaf, kapok, cocoa, hemp, henequen, sisal, jute, and ramie are representative instances of natural fibers acquired from plant sources. Animal-derived fibers such as silk, contrast hair, and wool. Plastic and thermosetting synthetic textile products, such as polyvinyl chloride, polyester, polyethylene terephthalate, polystyrene, high-density polyethylene, low-density polyethylene, viscid materials, and even compounds, generate waste subsequently their applications [29]. Primary and secondary recycling are used for final and post-consumer waste, respectively. Textile wastes in their end-of-life are frequently recycled through tertiary recycling, biodegradable, and incineration processes to create new products or energy recovery [110, 12]. Based on LCA, recycling process is commonly divided into four stages: primary, secondary, tertiary, and quaternary processing [88].

5.1 Primary Recycling

Primary textile waste can be simple or composite polymers that are easier to recycle. Post-consumer clothing waste denotes fiber materials that were discarded after being used. Textiles' biodegradability is defined by the origin of existence. This waste is typically used to manufacture clothing, rugs, belts, and composites [4, 44]. Primary regeneration, also known as CLR, is the method of creating a new textile product with desirable properties from waste [28]. Some additional processes, such as collecting as well as sorting of clothing materials, are required prior to recycling. Crushing, shredding, or milling processes are applied in the beginning for treating the textile wastes. The microfibers or threads are then blended with other polymer materials and necessary additives. The fundamental processing provides uniformity and refinement, as well as simplicity of creation [3]. [89] reported on the use of PES wastes blend with cotton fabrics to improve the mechanical properties by melt spinning process.

5.2 *Secondary Recycling*

Textiles having unspecified composition and purity are treated in the secondary recycling process. To improve purity, a few mechanical separating or purification treatments are used. This refining and separation usually necessitate the use of mineral acids, additives, and drying treatments, which reduce the mechanical characteristics of the textile products [2]. The secondary recycling process is governed by a variety of factors, including the nature of the secondary wastes, the constituents of the textile material, the purity quality of the final product, availability, additional types, cost, and processing techniques [13].

5.3 *Tertiary Recycling*

Chemical recycling is also called as feedstock recycle. Tertiary method involves chemical recycling (chemolysis) and thermolysis [50, 51, 117, 52]. The chemical method is used to break down polymers into different stages, such as oligomers or monomers, by reacting with specific chemical agents. Manufactured oligomers or monomers can be found for a variety of applications, including polymerization to recover identical polymers and fibers [39].

5.3.1 Cotton Blend of Polyester

Chemical recycling of CT, wool, and polyester are thought to be the main fiber products studied in other studies. First, cotton-containing wastes are frequently inspected for HMF [9, 85, 98] and bioethanol production [49]. Second, wool waste is often used in treatment with alkaline solutions such as nitrogen fertilizers or keratins. Wool is sometimes treated with high concentrations of ammonia at high temperatures to prevent excessive amino acid degradation [115]. Third, on a large scale, polyester fabrics and fiber products are used to recover terephthalate materials from waste. ([105], [108]). In the presence of polyester fiber, weak acid conditions are diagnosed using selected cotton fibers deformed with citric acid. Furthermore, in the presence of polyester, the wool fibers are selected and deformed by the weakly alkaline states created with sodium hydrogen carbonate. Figure 4 depicts the proposed chemical process for the waste treatment of polyester containing cotton and/or wool-based fiber products [56].

Cotton is the most probable primary source in various modern industries and increases the demand ratio. The present challenge is to find out other sources of cotton production with low cost and high quality, as agriculture produces too much cotton for these needs. The new approach consisted of three main processes. The first step is leakage using nitric acid, the second step is dissolution using dimethyl sulfoxide (DMSO), and the third one is bleaching with sodium hypochlorite (NaOCl)

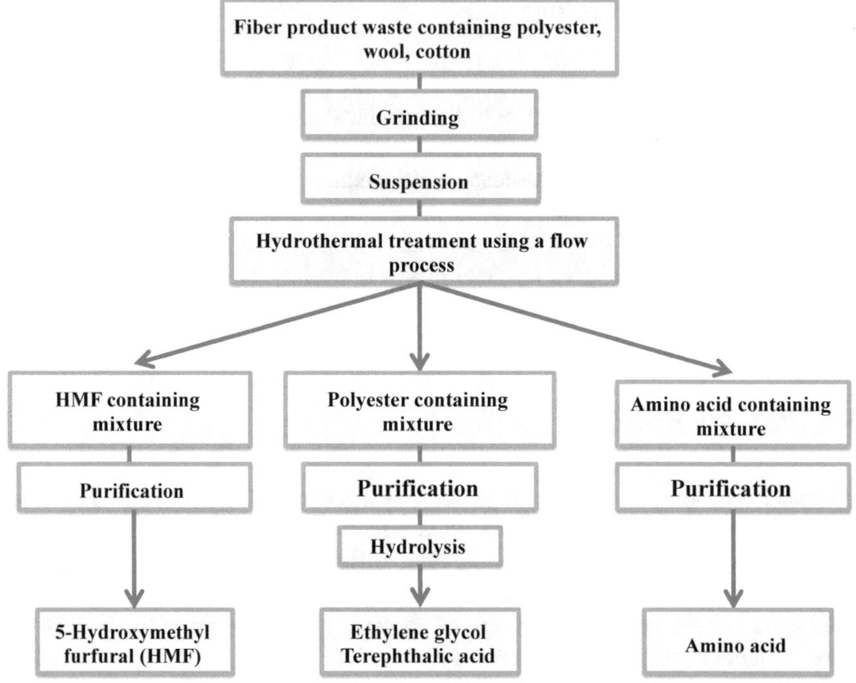

Fig. 4 Chemical recovery from fabric waste containing cotton, wool, and polyester fiber (*Credit* [56], Permission granted)

and dil. HCl, respectively. Pilot-scale experiments, the development strategy, help many to achieve self-sustenance in raw cotton fiber and accelerate the transition to vital economy; the carbon footprint of the recovery rate, economic performance, and technology are estimated as 93%, $ 1466/tons, and 1534 kg of CO_2-eq/tons [113].

5.3.2 Alkaline Hydrolysis

Cotton was hydrolyzed with hydrochloric acid to produce a microcrystalline cellulose powder that could be separated from the other polyester fabrics. The recovered polyester is re-rolled into PET yarns, which can then be used to make new textiles [100, 101]. Where polyester is maintained, the opposite, i.e., maintenance of cotton and degreasing or dissolving of polyethylene terephthalate (PET), may be on option. An approach that degrades polyester when maintaining cotton has been proposed using hydrolysis. Filtration was used to remove the polyester residue from 1 to 2% cellulose solution, which was made from cotton polyester blended compounds. Following that, the cellulose solutions are further enriched from 15 to 17% using cotton material from an alternate feed. The cellulosic component was spun from

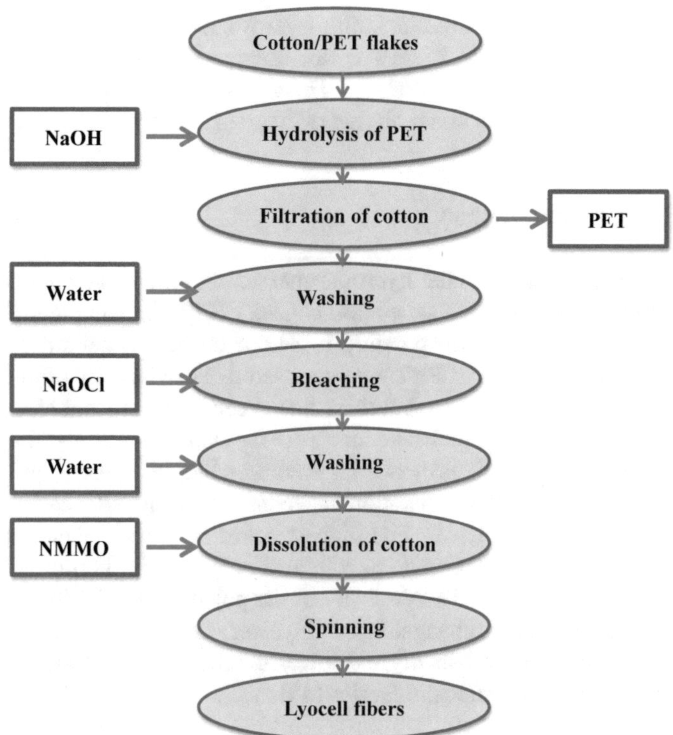

Fig. 5 The recycling of cotton from cotton/polyester fabrics as a New Lyocell Fiber *Credit* [65], Permission granted)

mesomorphic N-methyl morpholine N-oxide (NMMO) solutions like new lyocell fibers. This is a more environmentally friendly method of hydrolysis [65] (Fig. 5).

The 1-allyl-3-methylimidazolium chloride ([AMIM] [Cl]) and 1-butyl-3-methylimidazolium acetate ([BMIM] [OAc]) is illustrated to promote the recycling of mixed textile waste. This process algorithms can be compared with NMMO. Ionic liquids (ILs) dissolve the cellulose components, while filtration removes the polyester residue and recovers the subsequent freezing cellulose. Finally, the polyester (PES) component could be separated and recovered at higher yields [95]. The super based ionic liquid is using [DBNH] [OAc] is not required the addition of any stabilizers and allowing to adjust the operating conditions at a temperature of 30 °C lower than the spinning solutions prepared from NMMO. In the present patent, we are able to selectively dissolve cellulose in this solvent cotton-polyester composite, without severe degeneration of the polyester, and sufficient process conditions are used [41, 42].

The results of the investigation showed that a high degradation rate (95.47%) of the cotton components was achieved from 50% cotton to 50% polyester blended yarn waste. By dissolving cotton/polyester knitting yarns with a mixture of 200 g/1 $MgCl_2$ and 2 g/1 citric acid at 130 °C, 91.44% cotton component destruction was achieved with $MgCl_2$ and higher pre-concentrations. Treatment temperature at 20 °C compared with pretreatment with an aqueous solution of 20 g/1$MgCl_2$ and 4 $_g$/1$Al_2(SO_4)_3$ [87].

5.3.3 Hydrothermal Method

Using dilute HCl catalyzed under hydrothermal conditions, waste cotton/polyester blended fabrics achieve a good separation capacity (Fig. 6). The highest cellulose powder yield was 49.3%, corresponding to 84.5% change in cellulose in WBFs; additionally, more than 96% of PET were recovered. The viscosity of the polyester has decreased slightly, indicating that there have been no significant changes in the polyester after the hydrothermal reaction. This method is necessary for recycling composite fabrics because it provides an environmentally friendly and powerful method for separating WBFs. separation capacity, with dilute HCl catalyzed under hydrothermal conditions (Fig. 6). The highest cellulose powder yield was 49.3%, which is equivalent to 84.5% change in cellulose in WBFs; furthermore, more than 96% of PET were recovered. The viscosity of the polyester is slightly lower which does not indicate momentous changes in the polyester after the hydrothermal reaction. This method provides an eco-friendly and powerful process for separating WBFs and is essential for recycling composite fabrics [45].

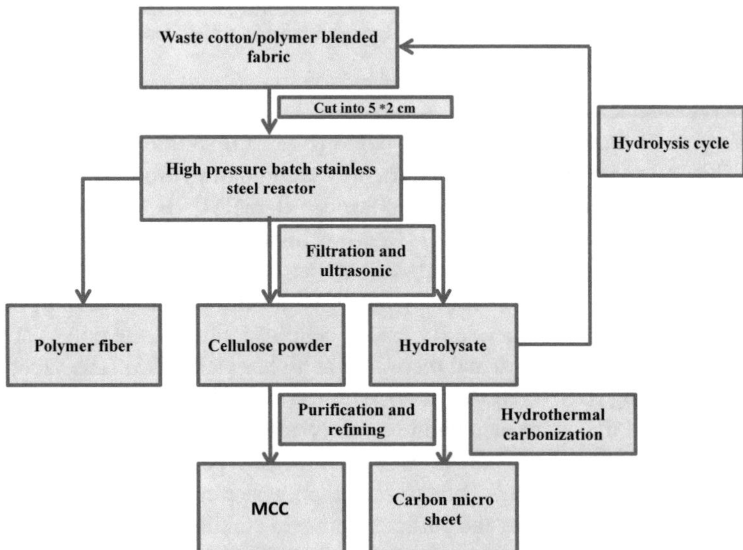

Fig. 6 Cotton and PET recovery from WBFs (*Credit*[45], Permission granted)

6 Biological Recycling of PES

Plastic biodegradation by microorganisms is a slow process, and it is affected by the environmental factors like temperature, pH, humidity, and ultraviolet rays. Biodegradation of PET usually causes shortening of the polymer chain which reduces the molecular weight of the polymer and subsequently, oligomers, dimers, water-soluble monomers are formed. The short polymer intermediates formed are assimilated by the microbial cell membrane as a carbon source to release CO_2. The products of biodegradation processes cannot be reused; however, it is a best eco-friendly alternative to incineration or landfill.

The enzymatic biodegradation of polyesters is a cost-effective and environmentally friendly approach. Enzymes depolymerize textile waste into their monomeric unit, which can be polymerized again to produce virgin quality polymers and fibers. Lipolytic enzymes that contain a large number of carboxyl esterases (EC 3.1.1.1) can hydrolyze various carboxylic esters into alcohols and organic acids as products. The most well-known esterases are α/β-hydrolases, which have a catalytic triad of serine, histidine, and aspartate as the active site. The free enzyme binds to the polyester substrate, and the hydroxyl group of serine forms a tetrahedral oxyanion intermediate by making a nucleophilic attack on the carbonyl carbon of the substrate. The oxyanion's negative charge is stabilized by an electrostatic interaction with the protein's > NH groups at a site known as the oxyanion hole. The tetrahedral intermediate then collapses to form an acyl enzyme complex, which is aided by a proton transfer from the positively charged histidine residue to the oxygen molecule. A second nucleophilic attack is produced by a nucleophile (water or alcohol in hydrolysis or esterification reactions, respectively) to the carbonyl carbon of the acyl enzyme complex to form the second tetrahedral intermediate. Then the intermediate is broken down to carboxylic acid product and the enzyme is regenerated [106]. All polyester degrading lipolytic (lipases, esterases, cutinases) enzymes and lipase-/cutinase-like (PETase, PHA depolymerases) enzymes and similar polyesterases catalyze polyester bond hydrolysis by the same mechanisms. The most significant studies related to the enzymatic degradation of PET are listed in Table 1.

6.1 Cutinases

Cutinases (EC 3.1.1.74) are serine esterases containing the classical Ser-His-Asp triad and belong to α/β-hydrolase superfamily. The cutinase was first identified and characterized from the filamentous fungus *Fusariumsolanipisi* [16]. Cutinases derived from *Fusarium solanipisi, Fusarium oxysporum, and Thermobifi da fusca* are commonly studied in enzymatic hydrolysis. Cutinase enzymes are multifunctional enzymes that can hydrolyze synthetic high molecular weight polyesters, short- and long-chain triacylglycerols, and soluble esters, as well as catalyze esterification and transesterification at 40–70°C and pH 7–9 [31].

Table 1 Enzymatic biodegradation of PET wastes

Microorganism	Enzyme	Inference	References
Actinomycete Thermobifida fusca	Hydrolase	Depolymerization of PET with the enzyme results in water-soluble oligomers or monomers that can be reused for synthesis	[61]
Thermobifida fusca (TfH)	Hydrolase	Aromatic polyesters with melting point above 200 °C can be enzymatically depolymerized. TfH is even capable of degrading commercial PET from beverage bottles	[60]
Thermobifida alba Est119	Esterases	It possesses the ability to significantly degrade aliphatic–aromatic co-polyester film	[46]
T.cellulosilytica DSM44535	Cutinase (Double mutant R29N and A30V; triple mutant R19S, R29N, and A30V)	The mutants strongly increased the hydrolysis activity for PET model substrate bis(benzoyloxyethyl) terephthalate and PET	[1]
Thermobifida alba	Cutinase	The enzyme released primarily 2-hydroxy ethyl benzoate from bis(benzoyloxyethyl)terephthalate (3PET)	[82]
Fusarium solani pisi	Cutinase	It degrades the PET film at a rate of 12 mg/h/mg of enzyme	[99]
Saccharomonospora viridis	Ca^{2+}-activated Cutinase	The calcium activated cutinase significantly degraded the PET films that were shown by weight loss and the amount of TPA produced	[55],
Fusarium oxysporum (FoCut5a)	Cutinase	FoCut5a can be used in biotechnological applications, such as the modification and degradation of PET-based fabrics and other synthetic materials	[22]
Triticum aestivum	Lipase	A significant decrease in the oligomer concentration was detected	[7]
Thermomyces lanuginosus	Lipase	A significant increase of hydrolysis products from the model substrate of PET was observed in the presence of Triton X-100 because of interfacial activation	[5]

(continued)

Table 1 (continued)

Microorganism	Enzyme	Inference	References
Thermobifida fusca and Fusarium solani	Cutinase	The addition of the plasticizer N,N-diethyl-2-phenylacetamide leads to enhanced hydrolysis rates for both lipase and cutinase on semicrystalline PET polymers (film and fabric)	[5]
Thermobifida fusca	Carboxylesterases	Cyclic PET trimers hydrolyzed with TfCa with optimal activity at 60°C and a pH of 6	[10]
Pseudomonas aestusnigri VGXO14	Polyester hydrolase	PET hydrolysis was detected at 30°C with an amorphous PET film, but not with a commercial PET bottle PET film	[11]
Thermobifida fusca KW3	Carboxylesterase	TfCa hydrolyzed the PET nanoparticles, 0.01 mg per mL of TfCa released 7 nmol of the PET hydrolysis product terephthalic acid per mL	[104]
Ideonella sakaiensis 201-F6	PETase & MHETase	The strain produces two enzymes capable of hydrolyzing PET and the reaction intermediate, mono(2-hydroxy ethyl) terephthalic acid. Both enzymes convert PET efficiently into its monomers, terephthalic acid, and ethylene glycol	[112]
Thermobifida fusca KW3	Cutinase TfCut2 Dual enzyme polyester hydrolase and immobilized carboxylesterase	Enzymatic PET hydrolysis is inhibited by the degradation intermediate mono-(2-hydroxyethyl) terephthalate (MHET). Continuous hydrolysis of the released intermediates MHET and BHET by the dual enzyme increased the degradation rate of PET films	[8]
Candida Antarctica lipase B (CALB) and *Humicola insolens* cutinase (HiC)	lipase and cutinase	A mixture of HiC and CALB improves the TPA titer during PET hydrolysis PET depolymerization increases by soaking the PET particles in MEG	[14]
Bacillus subtilis	PETase from the bacterium HR29	It is the most thermostable bacterial PETase that shows a melting temperature of 101 °C	[111]

(continued)

Table 1 (continued)

Microorganism	Enzyme	Inference	References
Bacillus subtilis	p-nitrobenzylesterase	The enzyme released TPA, benzoic acid (BA), 2-hydroxyethyl benzoate (HEB), and MHET from PET	[84]
Thermobifida halotolerans (Thh_Est)	esterase	Thh_Est improved hydrophilicity of polyester surfaces which significantly increases the hydrolysis	[81]
Thermomonospora curvata Tcur1278 and Tcur0390	Putative polyester hydrolases	Tcur1278 showed higher thermal stability and high hydrolytic activity on PET Tcur0390 showed higher hydrolytic activity against PCL and PET nanoparticles compared to Tcur1278	[109]
Fusarium oxysporum	Cutinase	The enzymatic modification can functionalize textile surface made of PET, without compromising the polymer bulk properties, such as strength	[118]
Thermobifida cellulosilytica (Thc_Cut1)	Cutinase	Thc_Cut1 selectively hydrolyzes PET moieties in packaging and bottles made of PET and polyethylene (PE) or polyamide (PA) blends, releasing TPA and MHET Polymer blends hydrolyzed 9 times higher compared to higher crystalline pure PET	[32]
Humicola insolens cutinase (HiC)	Cutinase	Chemical pretreatment carried out in an environmentally friendly way leads to depolymerization of the polyester-composed waste textiles, yielding approximately 85% TPA The enzymatic hydrolysis performed in a second reaction step leads to further hydrolysis of the remaining oligomers yielding TPA with a purity of 97%	[78]
Suberinase—genome of Streptomyces scabies	Protein Sub1	Sub1 activity on PET enhanced by the addition of Triton-producing terephthalic acid	[48]

(continued)

Table 1 (continued)

Microorganism	Enzyme	Inference	References
Pichia pastoris ATCC76273	LC-cutinase	LCC showed elevated thermostability because of glycosylation and also improved catalytic performance for PET hydrolysis	[92]
Ideonella sakaiensis strain 201-F6	Leaf–branch compost cutinase	The enzyme catalyzed PET depolymerization to 90% conversion in less than 10 h, with a mean productivity of 16.7gTAl^{-1} h^{-1} The resulting purified TPA monomers were used to synthesize PET	[107]
Thermobifida fusca Tfu_0883	Cutinase mutant I218A double mutant Q132A/T101A	Both single and double mutants exhibited considerably higher hydrolysis efficiency	[94]
Fusarium solani pisi	Cutinase CUT CUT-N1	Enzymatic treatment of PET by cutinases for 48 h decreases the mass of PET particles	[36]
Thermobifida cellulosilytica (Thc_Cut1)	Cutinase enzyme fused to the class II hydrophobins HFB4 and HFB7 and the pseudo-class I hydrophobin HFB9b	PET hydrolysis enhanced > 16-fold times compared to free enzyme	[83]

Cutinase from *F. solanipisi* was subjected to the hydrolysis of industrially important esters including synthetic esters and to degrade insoluble polymer films of poly(ethylene terephthalate) [120], poly(ε-caprolactone) (PCL) [119], and poly(butylene succinate) [69]. The neutral charge in the crowning area of the active site and the disulfide bond are attributed to the stability of the enzyme [66]. The reengineering of enzymes leads to improved function and stability over a range of temperature and pH.

The activity and stability of the enzymes are regulated by binding metal ions to them. Miyakawa et al. (55) found that Ca^{2+} unbound cutinases are inactive while Ca^{2+} bound cutinases are active. The binding of divalent ions stabilize the loop structure of the enzyme. The addition of Ca^{2+} or Mg^{2+} cations to the binding site of *Thermobifida fusca* increases the melting point and thermostability which are sufficient to degrade semi-crystalline PET films at 65°C [103]. *Thermobifida cellulosilytica cutinase* (Thc_Cut1) was examined for the hydrolysis of PET moieties in polymer blends. An increase in the crystalline nature of pure PET decreases the hydrolysis ability of cutinase [33].

The PET hydrolysis with wild-type cutinases was only capable of modifying the surface of the fibers, and the yields are very low. The mutations were designed to support the enlargement of the active site of the enzymes and enhance its hydrophobic character. The modified cutinase is expected to better accommodate the synthetic fiber, leading to an increase of hydrolysis ability. The recombinant cutinase from F. solani was designed without altering the amino acid sequence and overexpressed in P. pastoris. Cutinase cloned from *Fusariumsolani* degraded poly(butylene succinate) films 100% within 6 h [46] whereas only 80% degradation of PBS was achieved after 21 days of degradation with Fusarium sp. FS1301.

[31] studied the degradation of poly(ethylene terephthalate) with *Thermobifdafusca Cutinase* and mutant TfCut2 G62A/F209A modified with surfactant. The electrostatic attraction between the enzyme and the added surfactant enhances the binding of low-crystallinity PET film to the enzyme and later hydrolysis. The catalytic activity of the enzyme TfCut2 increased with the addition of cationic surfactant. The G62A/F209A double mutant showed 90 $\pm$ 4.5% degradation in the presence of cationic surfactant after 24 h. The double mutant showed 12.7 times higher activity than the wild-type TfCut2.

6.2 *Lipases*

Lipases are important industrial enzymes that catalyze the hydrolysis of water-insoluble triglycerides with long chains (greater than C10). The fold core domain of α/β-hydrolase is adjacent to the lid or flap domain, which covers the active site. The opening of the lid can create a large hydrophobic substrate binding groove, which helps hydrophobic polyesters stick better [37]. Therefore, the microbial lipases show lower activity toward PET and other aromatic polyesters due to their lid domain that covers the hydrophobic active site.

Lipases showed a significantly higher hydrolysis rate toward aliphatic polyester and were also reported to degrade PET fabrics to some extent by enhancing their wettability, dyeability, and absorbent characteristics [39]. [57], used *Candida antarctica* lipase as a catalyst in toluene at 60 °C to produce oligomers from poly(-caprolactone). By removing the solvent, the residual oligomer was polymerized again with the same catalyst. [15] used the assortment of lipase from *Candida Antarctica* (C. antarctica lipase lB CALB) and HiC for efficient PET hydrolysis to TPA. The combination of both enzymes synergistically enhanced the hydrolysis of PET.

6.3 *Esterase*

Esterases hydrolyze ester bonds, which causes microbial degradation of polyester-polyurethane (PUR) [63]. PURestersase from *Comamonas acidovorans* TB-35 has a hydrophobic PUR surface binding domain and a catalytic domain where the essential surface binding domain degrades PUR and releases diethylene glycol and adipic acid as biodegradation products.

Nocardia with the help of the esterase enzyme reported slow microbial degradation of PET [121]. Recombinant esterase from *ThermobifidaThh_Est* showed efficient surface hydrolysis for PET polyester to TPA and MHET, and its effect was equivalent to cutinases [82]. [55] used recombinant thermostabilized polyesterase from *Saccharo monosporaviridis* AHK190, which is capable of hydrolyzing PET, and found that the activity of PET hydrolysis increased in the presence of calcium ions.

6.4 PETase

PETase (3.1.1.101) was discovered by [112] from the bacterium I. sakaiensis 201-F6 which secretes two enzymes, PETase and MHETase. PETase first hydrolyzed the PET polymer in MHET, which was further hydrolyzed in TPA and EG by MHETase.

6.5 PHA Depolymerases and Other Polyesterases

PHB and PHBV hydrolyses are catalyzed by specific enzymes PHB and PHBV depolymerases, respectively, which have the same catalytic triad, lid domain, and hydrolysis mechanism as all serine hydrolases. Polyester hydrolases PHL7 [96] from a plant compost metagenome have been shown to degrade amorphous PET films and post-consumer PET thermoform packaging with high efficiency. The presence of leucine at position 210 in the enzyme PHL7 contributed to the binding of polymer substrate to the enzyme, resulting in high PET hydrolytic activity when compared to other polyester hydrolases, according to structural analysis. A dual enzyme system consisting of a polyester hydrolase and the immobilized carboxylesterase TfCa from *T. fusca* KW3 improved biocatalytic degradation of PET films. The HPLC analysis of the product of the double enzyme reaction reveals a higher proportion of soluble products and a lower proportion of MHET. The polyester hydrolase degraded the synthetic polymer chain, while the immobilized carboxylesterase hydrolyzed the intermediate MHET to TPA in the dual enzyme system [8].

7 Challenges Ahead

Major challenges faced using recycled PES: (1) Meeting the supply against demand and (2) Pure PES is separated from its blends with other fibers. Due to the technological issue with the recycling of PES is a major issue to provide sufficient supply. Challenges remaining in the recycled PES sector are to reduce the material's cost, improve material recovery as well as the quality of the recycled PES. Most recycling companies adopt to convert PET bottles into PES yarns through a melt recycling (physical) process. On the other hand, though the chemical recycling process provides pure monomers, the cost of the chemical recycling process becomes a challenge for textile industries for implementation. However, there are companies like Carbios, GR3N, Loop, Resinate, Worn Again, and Polygenta have taken initiatives to improve the chemical recycling process efficiency and reduction of cost [53]. On the other hand, the lack of an industrial process for the separation of PES from other fabrics such as cotton and wool are another challenge in the recycling of textile waste. However, chemical and biological methods for recycling PES blends have now been developed. [79] have explored the separation of wool and cotton from the blends of cotton-wool-PES blends using sequential extraction process: (1) using protease for separation of wool and (2) cellulases for the separation of cotton with 95 and 85% efficiency, respectively. [64] utilized an enzymatic method for the separation of wool and PES blends using keratinase enzyme to degrade wool fiber and leave PES in pure form.

Carbios, France, uses a patented biotechnology for recycling all types of PET waste through the enzymatic depolymerization of PET into terephthalic acid [107]. GR3N (Switzerland) uses microwave promoted alkaline hydrolysis of post-consumer PET bottles as well as PES textiles into terephthalic acid.

8 Future Perspectives and Conclusions

The textile industries have capacities to achieve circular economy through adopting pre- and post-consumer PES textile wastes. A road map must be needed for the collection, segregation, and recycling of post-consumer textile waste. Chemical and biological recycling of PES waste is an excellent solution for providing high-quality PES fibers rather than a mechanical or physical recycling process. Increasing the volume of PES recycling could substantially reduce microplastic pollution and conserve the use of traditional petrochemicals. There are challenges in chemical or biological recycling of waste PES blends that could possibly be overcome in near future. At any point of LCA, waste PES textiles could be reused and recycled to textile or for any other applications.

References

1. Acero EH, Ribitsch D, Dellacher A, Zitzenbacher S, Marold A, Steinkellner G, Gruber K, Schwab H, Guebitz GM (2013) Surface engineering of a cutinase from *Thermobifida cellulosilytica* for improved polyester hydrolysis. Biotechnol Bioeng 110:2581–2590. https://doi.org/10.1002/bit.24930
2. Achilias DS, Andriotics I, Kousidis IA, Louka DA, Nianias NP, Siafaka P (2012) Recent advances in the chemical recycling of polymers (PP, PS, LDPE, HDPE, PVC, PC, Nylon, and

PMMA. In Mateial Recycling—Trends and Perspective, In Tech Open
3. Al-Salem S, Lettieri P, Baeyens J (2010) Valorization of Plastic Solid Waste (PSW) by primary to quaternary routes from reuse to energy and chemicals. Prog Energy Combust Sci 36:103–129. https://doi.org/10.1016/j.pecs.2009.09.001
4. Andrady, A (2003) An environmental primer, plastic and the environment. In: Andrady, A (ed) Wiley Interscience, Hoboken, NY, USA
5. Anita E, Sonja H, Tina B, Rita A, Artur CP, Franz K, Wolfgang K, Georg MG (2009) Enzymatic surface hydrolysis of poly(ethylene terephthalate) and bis(benzoyloxyethyl) terephthalate by lipase and cutinase in the presence of surface active molecules. J Biotechnol 143:207–212. https://doi.org/10.1016/j.jbiotec.2009.07.008
6. Awaja F, Pavel D (2005) Recycling of PET. Eur Polymer J 41:1453–1477. https://doi.org/10.1016/j.eurpolymj.2005.02.005
7. Axel N, Axel B, Monika N, Marcus K, Axel K, Monika W, Matthias G (2006) A contribution to the investigation of Enzyme-Catalysed Hydrolysis of Poly(ethylene terephthalate) Oligomers. Macromol Mater Eng 291:1486–1494. https://doi.org/10.1002/mame.200600204
8. Barth M, Honak A, Oeser T, Wei R, Belisário-Ferrari MR, Then J, Schmidt J, Zimmermann W (2016) A dual enzyme system composed of a polyester hydrolase and a carboxylesterase enhances the biocatalytic degradation of polyethylene terephthalate films. Biotechnol J 11(8):1082–1087. https://doi.org/10.1002/biot.201600008
9. Bicker M, Hirth J, Vogel H (2003) Dehydration of fructose to 5-hydroxymethylfurfural in sub- and supercritical acetone. Green Chem 5:280–284. https://doi.org/10.1039/b211468b
10. Billig S, Oeser T, Birkemeyer C, Zimmermann W (2010) Hydrolysis of cyclic poly(ethylene terephthalate) trimers by a carboxylesterase from Thermobifida fusca KW3. Appl Microbiol Biotechnol 87(5):1753–1764. https://doi.org/10.1007/s00253-010-2635-y
11. Bollinger A, Thies S, Knieps-Grünhagen E, Gertzen C, Kobus S, Höppner A, Ferrer M, Gohlke H, Smits SHJ, Jaeger KE (2020) A novel Polyester Hydrolase from the Marine Bacterium Pseudomonas aestusnigri—Structural and functional insights. Front Microbiol 13(11):114. https://doi.org/10.3389/fmicb.2020.00114
12. Boustead I (2005) Ecoprofiles of the European Plastics Industry. Polyethylene terephthalate (PET) amorphous grade, Plastic Europe Report
13. Brems A, Baeyens J, Dewil R (2012) Recycling and recovery of post-consumer plastic solid waste in a European context. Therm Sci 16:669–685. https://doi.org/10.2298/tsci120111121b
14. Carniel A, Valoni É, Nicomedes J, da Conceição GA, de Castro A (2016) Lipase from *Candida antarctica* (CALB) and cutinase from *Humicola insolens* act synergistically for PET hydrolysis to terephthalic acid. Process Biochem 59A:84–90. https://doi.org/10.1016/j.procbio.2016.07.023
15. de Castro AM, Carniel A, Nicomedes Junior J, da Conceição GA, Valoni É (2017) Screening of commercial enzymes for poly(ethylene terephthalate) (PET) hydrolysis and synergy studies on different substrate sources. J Ind Microbiol Biotechnol 44(6):835–844. https://doi.org/10.1007/s10295-017-1942-z
16. Chen S, Su L, Chen J, Wu J (2013) Cutinase: Characteristics, preparation, and application. Biotechnol Adv 31(8):1754–1767. https://doi.org/10.1016/j.biotechadv.2013.09.005
17. Chen X, Memon HA, Wang Y, Marriam I, Tebyetekerwa M (2021) Circular economy and sustainability of the clothing and textile industry. Materials Circular Economy 3(1):12. https://doi.org/10.1007/s42824-021-00026-2
18. Cuc S, Vidovic M (2011) Environmental sustainability through clothing recycling. Operations And Supply Chain Management 4:108–115. https://doi.org/10.31387/oscm0100064
19. Cura K, Rintala N, Kamppuri T, Saarimäki E, Heikkilä P (2021) Textile recognition and sorting for recycling at an automated line using near infrared spectroscopy. Recycling 6(1):11. https://doi.org/10.3390/recycling6010011
20. Curran MA (2012) Life cycle assessment handbook: A guide for environmentally sustainable products. Wiley, Hoboken
21. Deepti G, Harshita C, Charu G (2015) Topographical changes in polyester after chemical, physical and enzymatic hydrolysis. The Journal of The Textile Institute 106:690–698. https://doi.org/10.1080/00405000.2014.934046

22. Dimarogona M, Nikolaivits E, Kanelli M, Christakopoulos P, Sandgren M, Topakas E (2015) Structural and functional studies of a Fusarium oxysporum cutinase with polyethylene terephthalate modification potential. Biochimica et Biophysica Acta (BBA): General Subjects 1850:2308–2317. https://doi.org/10.1016/j.bbagen.2015.08.009
23. Dissanayake DGK, Weerasinghe DU (2021) Fabric waste recycling: A systematic review of methods, applications, and challenges. Materials Circular Economy. 3:24. https://doi.org/10.1007/s42824-021-00042-2
24. Dobilaite V, Mileriene G, Juciene M, Saceviciene V (2017) Investigation of current state of pre-consumer textile waste generated at Lithuanian enterprises. International Journal of Clothing Science and Technology 29:491–503. https://doi.org/10.1108/IJCST-08-2016-0097
25. Dris R, Gasperi J, Saad M, Mirande C, Tassin B (2016) Synthetic fibers in atmospheric fallout: A source of microplastics in the environment. Mar Pollut Bull 104:290–293. https://doi.org/10.1016/j.marpolbul.2016.01.006
26. Dris R, Gasperi J, Mirande C, Mandin C, Guerrouache M, Langlois V, Tassin B (2017) A first overview of textile fibers, including microplastics, in the indoor and outdoor environment. Environ Pollut 221:453–458. https://doi.org/10.1016/j.envpol.2016.12.013
27. Echeverria CA, Handoko W, Pahlevani F, Sahajwalla V (2019) Cascading use of textile waste for the advancement of fibre reinforced composites for building applications. J Clean Prod 208:1524–1536. https://doi.org/10.1016/j.jclepro.2018.10.227
28. Esteve-Turrillas F, Guardia MDL (2017) Environmental impact of cotton recovery in the textile industry. Resour Conserv Recycl 116:107–115. https://doi.org/10.1016/j.resconrec.2016.09.034
29. Fangueiro R and Rana S (2016) Advances in science and technology towards industrial applications. In: Natural Fibres (eds) Springer, Netherlands
30. Fei X, Freeman HS, Hinks D (2020) Toward closed loop recycling of polyester fabric Step 1 decolorization using sodium formaldehyde sulfoxylate. J Clean Prod 254:120027
31. Furukawa M, Kawakami N, Tomizawa A, Miyamoto K, (2019) Efficient degradation of poly(ethylene terephthalate) with *Thermobifida fusca* Cutinase exhibiting improved catalytic activity generated using mutagenesis and additive-based approaches. Sci Rep 9:16038. https://doi.org/10.1038/s41598-019-52379-z
32. Gamerith C, Vastano M, Ghorbanpour SM, Zitzenbacher S, Ribitsch D, Zumstein MT, Sander M, Herrero Acero E, Pellis A, Guebitz GM (2017b) Enzymatic degradation of Aromatic and Aliphatic polyesters by P Pastoris expressed Cutinase 1 from *Thermobifida cellulosilytica*. Front Microbiol 24(8):938. https://doi.org/10.3389/fmicb.2017.00938
33. Gamerith C, Zartl B, Pellis A, Guillamot F, Marty A, Acero EH, Guebitz GM (2017a) Enzymatic recovery of polyester building blocks from polymer blends. Process Biochem 59:58–64. https://doi.org/10.1016/j.procbio.2017.01.004
34. Garcia JM, Robertson ML (2017) The future of plastics recycling. Science 358(6365):870–872. https://doi.org/10.1126/science.aaq0324
35. Gasperi J, Wright SL, Dris R, Collard F, Mandin C, Guerrouache M, Langlois V, Kelly FJ, Tassin B (2018) Microplastics in air: Are we breathing it in. Curr Opin Environ Sci Heal 1:1–5. https://doi.org/10.1016/j.coesh.2017.10.002
36. Gomes D, Matamá T, Cavaco-Paulo A, Takaki G, Salgueiro A (2013) Production of heterologous cutinases by E. coli and improved enzyme formulation for application on plastic degradation. Electronic Journal of Biotechnology 16(5). https://doi.org/10.2225/vol16-issue5-fulltext-12
37. Gricajeva A, Nadda AK, Gudiukaite R (2021) Insights into polyester plastic biodegradation by carboxyl ester hydrolases. Journal of Chemical Technology & Biotechnology Oneline version. https://doi.org/10.1002/jctb.6745
38. Gulich, B (2006) Development of products made of reclaimed fibers. In: Wang Y (ed) Recycling Textiles Wood head. London, pp 117–136
39. Gupta VB, Kothari VK (1997) Manufactured fibre. Technology. https://doi.org/10.1007/978-94-011-5854-1

40. Harmsen P, Scheffer M, Bos H, (2021) Textiles for circular fashion: The logic behind recycling options. Sustain 13. https://doi.org/10.3390/su13179714
41. Haslinger S, Hummel M, Sixta H (2017) Separation and upcycling of cellulose containing blended waste, 42pp, Patent Cooperation Treaty—PCT/F120 17/050916
42. Haslinger S, Hummel M, Anghelescu-Hakala A, Määttänen M, Sixta H, (2019) Upcycling of cotton polyester blended textile waste to new man-made cellulose fibers. Waste Manag 97:88–96. https://doi.org/10.1016/j.wasman.2019.07.040
43. Hatamlou M, Özgüney AT, Özdil N, Mengüç GS (2020) Performance of recycled PET and conventional PES fibers in case of water transport properties. Ind Textila 71:538–544. https://doi.org/10.35530/IT.071.06.1691
44. Hawley JM (2006) Digging for diamonds: A conceptual framework for understanding reclaimed textile products. Cloth Text Res J 24:262–275. https://doi.org/10.1177/0887302x06294626
45. Hou W, Ling C, Shi S, Yan Z, Zhang M, Zhang B, Dai J (2018) Separation and characterization of waste cotton/polyester blend fabric with Hydrothermal method. Fibers and Polymers 19:742–750. https://doi.org/10.1007/s12221-018-7735-9
46. Hu X, Thumarat U, Zhang X, Tang M, Kawai F (2010) Diversity of polyester-degrading bacteria in compost and molecular analysis of a thermoactive esterase from Thermobifida alba AHK119. Appl Microbiol Biotechnol 87:771–779. https://doi.org/10.1007/s00253-010-2555-x
47. JCFA (Japan Chemical Fibers Association) (2017) Calculated by JCFA on the basis of the statistics listed in fiber Organon
48. Jabloune R, Khalil M, Ben Moussa IE, Simao-Beaunoir AM, Lerat S, Brzezinski R, Beaulieu C (2020) Enzymatic degradation of p-Nitrophenyl Esters, Polyethylene Terephthalate, Cutin, and Suberin by Sub1, a Suberinase encoded by the plant Pathogen Streptomyces scabies. Microbes Environ 35(1):ME19086. https://doi.org/10.1264/jsme2.ME19086.
49. Jeihanipour A, Taherzadeh MJ (2009) Ethanol production from cotton-based waste textiles. Biores Technol 100:1007–1010. https://doi.org/10.1016/j.biortech.2008.07.020
50. Jeya G, Ilbeygi H, Radhakrishnan D, Sivamurugan V (2017) Glycolysis of post-consumer Poly(ethylene terephthalate) wastes Using Al, Fe and Zn Exchanged Kaolin Catalysts with Lewis Acidity. Advanced Porous Materials 5:128–136. https://doi.org/10.1166/apm.2017.1141
51. Jeya G, Anbarasu M, Dhanalakshmi R, Vinitha V, Sivamurugan V (2020) Depolymerization of Poly(ethylene terephthalate) wastes through Glycolysis using Lewis Acidic Bentonite Catalysts. Asian Journal of Chemistry 32:187–191. https://doi.org/10.14233/ajchem.2020.22387
52. Jeya G, Dhanalakshmi R, Anbarasu M, Vinitha V, Sivamurugan V (2022b) A short review on latest developments in catalytic depolymerization of Poly (ethylene terephathalate) wastes. J Indian Chem Soc 99:100291. https://doi.org/10.1016/j.jics.2021.100291
53. Joo E, Oh J-E (2019) Challenges facing recycled polyester. Textile World
54. Jönsson C, Wei R, Biundo A, Landberg J, Schwarz Bour L, Pezzotti F, Toca AM Jacques L, Bornscheuer UT, Syrén PO (2021) Biocatalysis in the recycling landscape for synthetic polymers and plastics towards circular textiles. Chem Sus Chem 5 14(19):4028–4040. https://doi.org/10.1002/cssc.202002666
55. Kawai F, Oda M, Tamashiro T, Waku T, Tanaka N, Yamamoto M, Mizushima H, Miyakawa T, Tanokura M (2014) A novel Ca^{2+}-activated, thermostabilized polyesterase capable of hydrolyzing polyethylene terephthalate from Saccharomonospora viridis AHK190. Appl Microbiol Biotechnol 98:10053–10064. https://doi.org/10.1007/s00253-014-5860-y
56. Kawamura K, Sako K, Ogata T, Tanabe K (2020) Environmentally friendly, hydrothermal treatment of mixed fabric wastes containing polyester, cotton, and wool fibers: Application for HMF production. Bioresource Technology Reports 11. https://doi.org/10.1016/j.biteb.2020.100478
57. Kobayashi S (2010) Lipase-catalyzed polyester synthesis—A green polymer chemistry. Proceedings of the Japan Academy. Series B, Physical and biological sciences 86(4):338–365. https://doi.org/10.2183/pjab.86.338

58. Langley KDK, YK, (2006) Manufacturing nonwovens and other products using recycled fibers containing spandex. In: Wang Y (ed) Recycling in textiles. Woodhead, London, pp 137–214
59. Lorenz, P. http://thecircularlaboratory.com/recycled-polyester-and-the-plastic-bottle-dilemma
60. Mueller RJ (2006) Biological degradation of synthetic polyesters—Enzymes as potential catalysts for polyester recycling. Process Biochem 41:2124–2128. https://doi.org/10.1016/j.procbio.2006.05.018
61. Muller RJ, Schrader H, Profe J, Dresler K, Deckwer WD (2005) Enzymatic degradation of poly (ethylene terephthalate): rapid hydrolyse using a hydrolase from T. fusca. Macromol Rapid Commun 26:1400–1405. https://doi.org/1002/marc.200500410
62. Munasinghe P, Druckman A, Dissanayake DGK (2021) A systematic review of the life cycle inventory of clothing. J Clean Prod 320. https://doi.org/10.1016/j.jclepro.2021.128852
63. Nakajima-Kambe T, Shigeno-Akutsu Y, Nomura N, Onuma F, Nakahara T (1999) Microbial degradation of polyurethane, polyester polyurethanes and polyether polyurethanes. Appl Microbiol Biotechnol 51(2):134–140. https://doi.org/10.1007/s002530051373
64. Navone L, Moffitt K, Hansen KA, Blinco J, Payne A, Speight R (2020) Closing the textile loop: Enzymatic fibre separation and recycling of wool/polyester fabric blends. Waste Manag 1(102):149–160. https://doi.org/10.1016/j.wasman.2019.10.026
65. Negulescu II, Kwon H, Collier BJ, Collier JR, Pendse A (1998) Recycling cotton from cotton/polyester fabrics. Textile Chem, color, pp 30, 31–35
66. Nikolaivits E, Kanelli M, Dimarogona M, Topakas E (2018) A middle-aged Enzyme Still in its prime: Recent advances in the field of Cutinases. Catalysts 8. https://doi.org/10.3390/catal8120612
67. PET Recycling (2020) www.recycling-magazine.com/2020/06/08/pet-recycling-towards-a-circular-economy
68. Palacios-Mateo C, van der Meer Y, Seide G (2021) Analysis of the polyester clothing value chain to identify key intervention points for sustainability. Environ Sci Eur 33. https://doi.org/10.1186/s12302-020-00447-x
69. Pan W, Bai Z, Su T, Wang Z (2018) Enzymatic degradation of poly(butylene succinate) with different molecular weights by cutinase. Int J Biol Macromol 111:1040–1046. https://doi.org/10.1016/j.ijbiomac.2018.01.107
70. Park SH, Kim SH (2014) Poly (ethylene terephthalate) recycling for high value-added textiles. Fashion and Textiles 1(1). https://doi.org/10.1186/s40691-014-0001-x
71. Patti A, Cicala G, Acierno D (2020) Eco-sustainability of the textile production: Waste recovery and current recycling in the composites world. Polymers (Basel) 30:13(1):134. https://doi.org/10.3390/polym13010134
72. Payne A (2015) Open- and closed-loop recycling of textile and apparel products. Handbook of Life Cycle Assessment (LCA) of Textiles and Clothing 103–123. https://doi.org/10.1016/b978-0-08-100169-1.00006-x
73. Pensupa N, Leu SY, Hu Y, Du C, Liu H, Jing H, Wang H, Lin CSK (2017) Recent trends in sustainable textile waste recycling methods: Current situation and future prospects. Top Curr Chem (Cham) 16;375(5):76. https://doi.org/10.1007/s41061-017-0165-0
74. Peters RJ, Undas AK, Leeuwen SV (2020). Evaluation of the presence of potential hazardous substances from plastic and textile fibre recycling. https://doi.org/10.18174/515071
75. Piribauer B, Bartl A (2019) Textile recycling processes, state of the art and current developments: A mini review. Waste Manag Res 37(2):112–119. https://doi.org/10.1177/0734242X18819277
76. Piribauer B, Bartl A, Ipsmiller W (2021) Enzymatic textile recycling—Best practices and outlook. Waste Manag Res 39(10):1277–1290. https://doi.org/10.1177/0734242X211029167
77. Prata JC (2018) Airborne microplastics: Consequences to human health? Environ Pollut 234:115–126. https://doi.org/10.1016/j.envpol.2017.11.043
78. Quartinello F, Vajnhandl S, Volmajer Valh J, Farmer TJ, Vončina B, Lobnik A, Herrero Acero E, Pellis A, Guebitz GM (2017) Synergistic chemo-enzymatic hydrolysis of poly(ethylene

terephthalate) from textile waste. Microb Biotechnol 10(6):1376–1383. https://doi.org/10.1111/1751-7915.12734
79. Quartinello F, Vecchiato S, Weinberger S, Kremenser K, Skopek L, Pellis A, Guebitz GM (2018) Highly selective Enzymatic recovery of building blocks from Wool-Cotton-Polyester textile waste blends. Polymers (Basel) 7;10(10):1107. https://doi.org/10.3390/polym10101107
80. Riba JR, Cantero R, Canals T, Puig R (2020) Circular economy of post-consumer textile waste: Classification through infrared spectroscopy. J Clean Prod 272. https://doi.org/10.1016/j.jclepro.2020.123011
81. Ribitsch D, Acero EH, Greimel K, Dellacher A, Zitzenbacher S, Marold A, Rodriguez RD, Steinkellner G, Gruber K, Schwab H, Guebitz GM (2012b) A New Esterase from Thermobifida halotolerans Hydrolyses Polyethylene Terephthalate (PET) and Polylactic Acid (PLA). Polymers 4(4):617–629. https://doi.org/10.3390/polym4010617
82. Ribitsch D, Acero EH, Greimel K, Eiteljoerg I, Trotscha E, Freddi G, Schwab H, Guebitz GM (2012a) Characterization of a new cutinase from *Thermobifida alba* for PET-surface hydrolysis. Biocatal Biotransform 30(1):2–9. https://doi.org/10.3109/10242422.2012.644435
83. Ribitsch D, Herrero Acero E, Przylucka A, Zitzenbacher S, Marold A, Gamerith C, Tscheließnig R, Jungbauer A, Rennhofer H, Lichtenegger H, Amenitsch H, Bonazza K, Kubicek CP, Druzhinina IS, Guebitz GM (2015) Enhanced cutinase-catalyzed hydrolysis of polyethylene terephthalate by covalent fusion to hydrophobins. Appl Environ Microbiol 81(11):3586–3592. https://doi.org/10.1128/AEM.04111-14
84. Ribitsch D, Heumann S, Trotscha E, Herrero Acero E, Greimel K, Leber R, Birner-Gruenberger R, Deller S, Eiteljoerg I, Remler P, Weber T, Siegert P, Maurer KH, Donelli I, Freddi G, Schwab H, Guebitz GM (2011) Hydrolysis of polyethyleneterephthalate by p-nitrobenzylesterase from Bacillus subtilis. Biotechnol Prog 27(4):951–960. https://doi.org/10.1002/btpr.610
85. Rosatella AA, Simeonov SP, Frade RFM, Afonso CAM (2011) 5-Hydroxymethylfurfural (HMF) as a building block platform: Biological properties, synthesis and synthetic applications. Green Chem 13:54. https://doi.org/10.1039/c0gc00401d
86. Sandin G, Peters GM (2018) Environmental impact of textile reuse and recycling—A review. J Clean Prod 184:353–365. https://doi.org/10.1016/j.jclepro.2018.02.266
87. Sankauskaitė A, Stygienė L, Tumėnienė MD, Krauledas S, Jovaišienė L, Puodžiūnienė R (2014) Investigation of cotton component destruction in cotton/polyester blended textile waste materials. Mater Sci 20. https://doi.org/10.5755/j01.ms.20.2.3115
88. Scheirs J (1998) Polymer recycling, science and technology and applications. In: J. Wiley & Sons Chichester (ed) UK
89. Sharma K, Khilari V, Chaudhary BU, Jogi AB, Pandit AB, Kale RD (2020) Cotton based composite fabric reinforced with waste polyester fibers for improved mechanical properties. Waste Manag 15(107):227–234. https://doi.org/10.1016/j.wasman.2020.04.011
90. Shen L, Worrell E, Patel MK (2010) Open-loop recycling: A LCA case study of PET bottle-to-fibre recycling. Resour Conserv Recycl 55:34–52. https://doi.org/10.1016/j.resconrec.2010.06.014
91. Shen L, Worrell E, Patel MK (2012) Comparing life cycle energy and GHG emissions of bio-based PET, recycled PET, PLA, and man-made cellulosics. Biofuels, Bioprod Biorefn 6(6):625–639. https://doi.org/10.1002/bbb.1368
92. Shirke AN, White C, Englaender JA, Zwarycz A, Butterfoss GL, Linhardt RJ, Gross RA (2018) Stabilizing Leaf and Branch Compost Cutinase (LCC) with Glycosylation: Mechanism and effect on PET hydrolysis. Biochemistry 20;57(7):1190–1200. https://doi.org/10.1021/acs.biochem.7b01189
93. Shirvanimoghaddam K, Motamed B, Ramakrishna S, Naebe M (2020) Death by waste: Fashion and textile circular economy case. Sci Total Environ 20(718). https://doi.org/10.1016/j.scitotenv.2020.137317
94. Silva C, Da S, Silva N, Matamá T, Araújo R, Martins M, Chen S, Chen J, Wu J, Casal M, Cavaco-Paulo A (2011) Engineered Thermobifida fusca cutinase with increased activity on polyester substrates. Biotechnol J 6(10):1230–1239. https://doi.org/10.1002/biot.201000391

95. Silva RD, Wang X, Byrne N (2014) Recycling textiles: the use of ionic liquids in the separation of cotton polyester blends. RSC Adv 4:29094–29098. https://doi.org/10.1039/c4ra04306e
96. Sonnendecker C, Oeser J, Richter PK, Hille P, Zhao Z, Fischer C, Lippold H, Blázquez-Sánchez P, Engelberger F, Ramírez-Sarmiento CA, Oeser T, Lihanova Y, Frank R, Jahnke HG, Billig S, Abel B, Sträter N, Matysik J, Zimmermann W (2021) Low Carbon footprint recycling of post-consumer PET plastic with a Metagenomic polyester hydrolase. Chem Sus Chem 15. https://doi.org/10.1002/cssc.202101062
97. Stanescu MD (2021) State of the art of post-consumer textile waste upcycling to reach the zero waste milestone. Environ Sci Pollut Res Int 28(12):14253–14270. https://doi.org/10.1007/s11356-021-12416-9
98. Su Y, Brown HM, Huang X, Zhou XD, Amonette JE, Zhang ZC (2009) Single-step conversion of cellulose to 5-hydroxymethylfurfural (HMF), a versatile platform chemical. Appl Catal A 361:117–122. https://doi.org/10.1016/j.apcata.2009.04.002
99. Sulaiman S, Yamato S, Kanaya E, Kim JJ, Koga Y, Takano K, Kanaya S (2012) Isolation of a novel Cutinase Homolog with Polyethylene Terephthalate—degrading activity from Leaf-Branch compost by using a metagenomic approach. Appl Environ Microbiol 78:1556–1562. https://doi.org/10.1128/AEM.06725-11
100. Sun CH, Chen XP, Zhuo Q, Zhou T (2018a) Recycling and depolymerization of waste polyethylene terephthalate bottles by alcohol alkali hydrolysis. Journal of Central South University 25:543–549. https://doi.org/10.1007/s11771-018-3759-y
101. Sun H, Li C, Sheng C, Wang J, Deng, H, Jia, R, Xia, Y, Tian, F, Lu, Q, Guo, D, (2018b) Method for separating and preparing microcrystalline cellulose from polyester-cotton fabric, pp 9
102. Terazono A, Yoshida A, Yang J, Moriguchi Y, Sakai SI (2004) Material cycles in Asia: especially the recycling loop between Japan and China. J Mater Cycles Waste Manage 6. https://doi.org/10.1007/s10163-004-0115-0
103. Then J, Wei R, Oeser T, Barth M, Belisário-Ferrari MR, Schmidt J, Zimmermann W (2015) Ca^{2+} and Mg^{2+} binding site engineering increases the degradation of polyethylene terephthalate films by polyester hydrolases from Thermobifida fusca. Biotechnol J 10(4):592–598. https://doi.org/10.1002/biot.201400620
104. Thorsten O, Ren W, Thomas B, Susan B, Christina F, Wolfgang Z (2010) High level expression of a hydrophobic poly(ethylene terephthalate)-hydrolyzing carboxylesterase from Thermobifida fusca KW3 in Escherichia coli BL21(DE3). J Biotechnol 146:100–104. https://doi.org/10.1016/j.jbiotec.2010.02.006
105. Tokoro M (2010) A new step forward in clothing recycling. Material Cycles and Waste Management Research 21:157–168
106. Toledo M V, Llerena Suster CR, Ferreira ML, Collins SE, Briand LE, (2017) Molecular recognition of an acyl–enzyme intermediate on the lipase B from Candida antarctica. Catal Sci Technol 7:1953–1964. https://doi.org/10.1039/C7CY00245A
107. Tournier V, Topham CM, Gilles A, David B, Folgoas C, Moya-Leclair E, Kamionka E, Desrousseaux ML, Texier H, Gavalda S, Cot M, Guémard E, Dalibey M, Nomme J, Cioci G, Barbe S, Chateau M, André I, Duquesne S, Marty A (2020) An engineered PET depolymerase to break down and recycle plastic bottles. Nature 580:216–219. https://doi.org/10.1038/s41586-020-2149-4
108. Wang Y (2010) Fiber and textile waste utilization. Waste and Biomass Valorization 1:135–143. https://doi.org/10.1007/s12649-009-9005-y
109. Wei R, Oeser T, Then J, Kühn N, Barth M, Schmidt J, Zimmermann W (2014) Functional characterization and structural modeling of synthetic polyester-degrading hydrolases from Thermomonospora curvata. AMB Express 3(4):44. https://doi.org/10.1186/s13568-014-0044-9
110. Woolridge AC, Ward GD, Phillips PS, Collins M, Gandy S (2006) Life cycle assessment for reuse/recycling of donated waste textiles compared to use of virgin material: An UK energy saving perspective. Resour Conserv Recycl 46:94–103. https://doi.org/10.1016/j.resconrec.2005.06.006

111. Xi X, Ni K, Hao H, Shang Y, Zhao B, Qian Z (2021) Secretory expression in Bacillus subtilis and biochemical characterization of a highly thermostable polyethylene terephthalate hydrolase from bacterium HR29. Enzyme Microb Technol 143. https://doi.org/10.1016/j.enzmictec.2020.109715
112. Yoshida S, Hiraga K, Takehana T, Taniguchi I, Yamaji H, Maeda Y, Toyohara K, Miyamoto K, Kimura Y, Oda K (2016) A bacterium that degrades and assimilates poly(ethylene terephthalate). Science 11;351(6278):1196–9. https://doi.org/10.1126/science.aad6359
113. Yousef S, Tatariants M, Tichonovas M, Sarwar Z, Jonuškienė I, Kliucininkas L (2019) A new strategy for using textile waste as a sustainable source of recovered cotton. Resour Conserv Recycl 145:359–369. https://doi.org/10.1016/j.resconrec.2019.02.031
114. Zimmermann W (2020) Biocatalytic recycling of polyethylene terephthalate plastic. Philos Trans A Math Phys Eng Sci 24;378(2176):20190273 https://doi.org/10.1098/rsta.2019.0273
115. Zoccola M, Aluigi A, Tonin C (2009) Characterisation of keratin biomass from butchery and wool industry wastes. J Mol Struct 938:35–40. https://doi.org/10.1016/j.molstruc.2009.08.036
116. Zou Y, Reddy N, Yang Y (2011) Reusing polyester/cotton blend fabrics for composites. Composites B: Engineering 42:763–770. https://doi.org/10.1016/j.compositesb.2011.01.022
117. Jeya G, Rajalakshmi S, Gayathri KV, Priya P, Sakthivel P, Sivamurugan V, (2022a) A Bird's eye view on sustainable management solutions for non-degradable plastic wastes. In: Vasanthy M, Sivasankar V, Sunitha TG (eds) Organic pollutants: Toxicity and solutions. Springer International Publishing, Cham, pp 503–534
118. Kanelli M, Vasilakos S, Nikolaivits E, Ladas S, Christakopoulos P, Topakas E, (2015) Surface modification of poly(ethylene terephthalate) (PET) fibers by a cutinase from Fusarium oxysporum. Process Biochem 50:1885–1892. https://doi.org/10.1016/j.procbio.2015.08.013
119. Liu M, Zhang T, Long L, Zhang R, Ding S, (2019) Efficient enzymatic degradation of poly (ε-caprolactone) by an engineered bifunctional lipase-cutinase. Polym Degrad Stab 160:120–125. https://doi.org/10.1016/j.polymdegradstab.2018.12.020
120. Ronkvist ÅM, Xie W, Lu W, Gross RA, (2009) Cutinase-Catalyzed Hydrolysis of Poly(ethylene terephthalate). Macromolecules 42:5128–5138. https://doi.org/10.1021/ma9005318
121. Sharon C, Sharon M, (2017) Studies on Biodegradation of Polyethylene terephthalate: A synthetic polymer. J Microbiol Biotechnol Res 2:248–257
122. Vogt BD, Stokes KK, Kumar SK, (2021) Why is recycling of post-consumer plastics so challenging? ACS Appl Polym Mater 3:4325–4346. https://doi.org/10.1021/acsapm.1c00648

Converting Textile Waste into Designer Wall and Floor Tiles: A New Approach to Recycle Textile Waste

Nidhi Sisodia and M. S. Parmar

Abstract Tiles are being used for the floor, wall, and roof. These are often made from ceramic. Ceramic tile production comprehends many different processes according to each different finished product. There are six steps involved in the manufacturing process of ceramic tiles. These are (i) clay preparation, either by dry grinding or by wet milling and atomization; (ii) forming or molding of the tile by either dry pressing or extrusion; (iii) glaze preparation; (iv) drying, glazing, and decoration of the tile; (v) kiln firing; and (vi) classification and packing. However, other materials such as clay, glass, and concrete are also being used. Nowadays, synthetic tiles are also being made using plastic waste material along with resins. Disposing of textile waste is a very big problem. Textile waste covers wide varieties of fibers including natural and synthetic. It is estimated that around 92 million tons of clothing waste are thrown out every year globally. Most of the portion goes in landfills. Nowadays, textiles are also produced using unconventional fibers such as cornhusk, rice and wheat straw, and sugarcane. Reconverting waste clothing material by mixing it with resins, natural and synthetic can be explored for developing high-quality products like wall and floor tiles. Keeping this in mind, in this study, various types of textile wastes like denim, cornhusk, and rice straw will be used to develop wall and floor tiles material having water and flame retardant properties.

Keywords Flame retardant · Resin · Textile waste · Wall & Floor tiles · Water resistance

1 Introduction

In recent years, global textile manufacturing has been steadily expanding. As a natural result of basic needs, global population growth and growing living standards have

N. Sisodia (✉) · M. S. Parmar
Northern India Textile Research Association, Sector-23, Rajnagar, Ghaziabad, India

M. S. Parmar
e-mail: drmsparmar@nitratextile.org

S. S. Muthu (ed.), *Sustainable Approaches in Textiles and Fashion*, Sustainable Textiles: Production, Processing, Manufacturing & Chemistry,
https://doi.org/10.1007/978-981-19-0530-8_7

led to an increase in textile demand, as well as overconsumption as a result of rapid fashion trends. According to a World Bank analysis, global municipal solid waste will increase by 70% by 2025 [1].

Textile waste disposal is a major issue. Textile waste includes a wide range of natural and synthetic fibers (Fig. 1). Every year, it is estimated that 92 million tons of garment waste are discarded worldwide. The majority of it ends up in landfills. Besides, there are some unconventional fibers that are also being used in textiles. Some of them are cornhusk, sugarcane, rice straw, and wheat straw. In agriculture, these fibers are considered agricultural waste. Agricultural waste is waste that is generated as a result of different agricultural processes [2]. Textile trimmings are commonly discarded as a waste product, which becomes an environmental hazard when they are burned in heaps, releasing highly toxic gases into the atmosphere. As a result, there is a requirement to turn them into useable materials that accomplish two goals: waste reduction and the creation of a new product [3].

Textile is the second most polluting industry in the world. Because a garment's average life period is three years, textiles generate a significant amount of waste. Dumped textile waste accounts for 5% of all landfill space worldwide. A large volume of fibrous waste from the textile industry and post-consumer products is discarded globally [4].

This is not only an environmental concern, but it is also a waste of valuable resources. Textile trimmings are commonly discarded as a waste product, which becomes an environmental hazard when they are burned in heaps, releasing highly toxic gases into the atmosphere. As a result, there is a requirement to It serves a dual purpose by converting them into usable materials [5].

Fig. 1 Textile waste

About 15% of fabric used in garment production is cut, thrown, and wasted on average. Similarly, it is estimated that over one million tons of textiles are discarded each year, with the majority of this coming from domestic sources [6].

Some of the waste fabric material is being used for making value-added products by reprocessing and converting it into fibrous material for making shoddy yarn. This shoddy yarn is ultimately used in making carpets and rugs. The waste or used woolen clothes are recycled and have been re-used in making a blanket and other useful products. Industries have successfully created products from recycled fibers such as carpets, cushions, clean-up products, home insulations, fiber stuffing clean-up products, mattress, pads, geotextiles, landscaping, and concrete reinforcement [6]. The waste fabric material is also reprocessed and converted into shoddy yarn to create value-added products. This low-quality yarn is eventually utilized to make carpets and rugs. Waste or old woolen fabrics are recycled and repurposed to make blankets and other useful items. Industries have made items from recycled fibers with great success [7].

Tiles are commonly utilized as flooring in kitchens, baths, parking lots, and rooftops, as well as dining room tabletops. Ceramic, porcelain, glass, stone, and metal are all used to make tiles. There has been no systematic investigation into the use of waste textile materials in the development of roof and wall tiles. Textile waste materials, such as low-grade cotton, are used in this research [8]. Three types of textile waste are evaluated: pre-consumer textile waste, post-consumer textile waste, and industrial textile waste. Pre-consumer textile waste includes scraps, damaged or defective material samples, fabric selvages, and leftover fabric from the cutting process (Fig. 2). About 15% of fabric used in garment production is cut, thrown, and wasted on average. The term "post-consumer textile waste" refers to household items or clothes that the owner no longer needs [9].

Fig. 2 Denim cutting

100% degradable materials such as cotton, linen, silk, and hemp are composted or upcycled into a value-added product. Synthetic fabrics can be upcycled as composites and building blocks for use in construction and soundproofing [10]. Traditionally, old textiles were recycled to be used as a mop or washcloth in the home, but with the present invasion of throwaway textiles, the use and throw strategy has become popular, and old textiles are discarded. Industrial textile waste is made up of textiles that have been used in industrial applications [11]. Conveyor belts, filters, geotextiles, and wiping rags are examples of industrial textile waste generated by industrial uses [12]. Polymer composites are made up of a range of short or continuous fibers that are held together by an organic polymer matrix. High tensile strength, high stiffness, high fracture toughness, good abrasion resistance, good puncture resistance, good corrosion resistance, and low cost are all advantages of polymer matrix composites [12].

Due to their similar qualities to designed fibers, cellulose fibers have recently been employed as a fiber-reinforcement material, particularly in conjunction with polymers in fiber-reinforced composites [13]. Every year, a considerable volume of fibrous trash is disposed of in landfills. This not only causes economic and environmental problems for society but also wastes resources. Although it is ideal to recycle fiber waste into the same goods, due to the processing labor, energy, and pollutants involved, this strategy may not be viable for many types of waste materials, particularly for organic waste [14]. Textile recycling has a long history in many parts of the world, and recently, many commercial textile and carpet recycling operations are emerging with various records of success [14]. Textile waste is one of the most polluting types of waste. Around 80 billion garments are produced yearly, resulting in over 1.3 billion tons of fabric waste, and nearly 75% of this trash is abandoned untreated, potentially ending up in landfills or being burned [15]. Currently, a substantial amount of waste textiles is disposed of by burning and burial, both of which pollute the environment. As a result, recycling textile waste into high-mechanical products via an environmentally acceptable approach is a pressing concern [16]. In one of the research waste cotton fibers obtained from Blue-Jeans are mixed with polyester concrete to improve upon their compressive strength [17]. In this work, three types of materials: corn husk fiber, wheat straw fiber, and shredded denim waste are taken for making designer wall and floor tiles using suitable matrix materials. Developed tiles were evaluated for mechanical and chemical properties.

2 Materials and Methods

2.1 Reinforcement Material and Other Chemicals

Cornhusk and rice straw waste along denim waste fabric is used for the study as reinforcement material sources from the industry as well as from local farmers. Water repellent and flame retardant chemical sourced from NITRA.

2.2 *Matrix Materials*

Fibers by themselves are unable to transmit the weight to one another, necessitating the use of a binding substance. In the case of composites, the matrix performs this function. The matrix that is employed has a significant impact on the mechanical properties of composites. Matrix is made up of resin and a hardener that are mixed. For the current experiment, epoxy resin and natural resin are used. The epoxy resin B-11 Bisphenol-A (epichlorohydrin) and hardener were obtained from Kailashpati Polymers Ghaziabad, Uttar Pradesh, India, and natural resin procured from Ghaziabad M/s Jai Laxmi chemicals, Ghaziabad, Uttar Pradesh, India.

2.3 *The Experimental Calculation for Fiber and Matrix Weight Fractions*

Fiber and matrix weight fractions were calculated for each category sample. Matrix was prepared by mixing well 75% by weight of epoxy resin and 25% by weight of hardener by volume fractions, respectively. Following is the calculation:

1. Amount of epoxy resin = 16.75 ml
2. Hardener = 8.25 ml

Total matrix amount = Amount of epoxy resin (i) + Amount of hardener(ii) = 25 ml

So, approximate weight of matrix used (Wm) = 25 gm

Weight of fiber (Wf) = 10 gm

So, composite weight (WC) = wt. of fiber (wf) + wt. of the matrix (wm)

WC = Wm + Wf = 35 gm

So, the weight fraction of fiber = wt. of fiber/wt. of composite

WF = wf/WC

WF = 10/35 100 = 28.5%

So weight fraction of matrix will be = wt. of matrix/wt. of composite

WM = wm/WC

WF = 25/35 100 = 17.5%

2.4 *Preparation of Composite Tile Material*

The composites of various thicknesses (8 mm, 10 mm, and 12 mm) were prepared by mixing denim waste, corn husk and rice straw waste with epoxy resin using a hot pressing machine. Developed tiles were evaluated using the standard test method IS 13630 (Parts 1 to 15). The conceptual framework is given in Fig. 3.

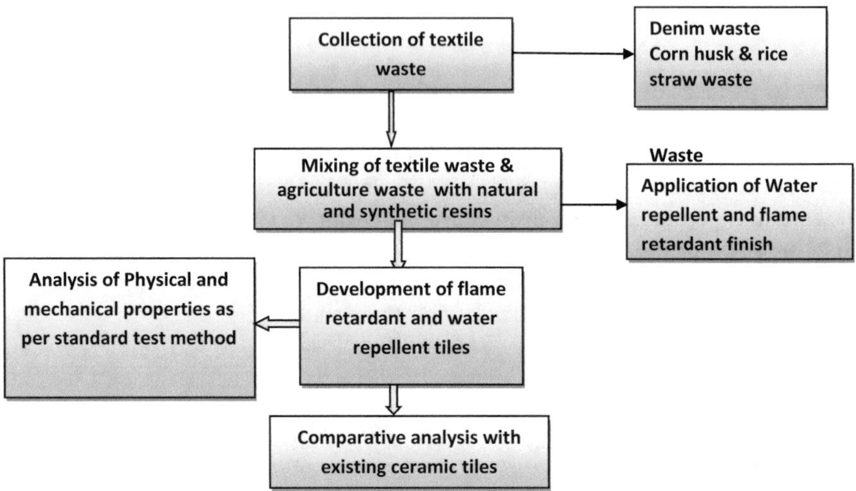

Fig. 3 Conceptual framework of the tile study

In the case of cornhusk and rice straw, extracted fibers were carded one time on carding machine to obtain a smooth web. The web was cut into 10 × 12 inch size mats. 2–3 of these mats were placed on top of each other based on the desired thickness layered composite. Natural resin and synthetic resin and hardener were mixed in different required ratios and applied on each mat one by one. Mats were sandwiched between two aluminum/Teflon sheets (having applied silicon oil) and set to cut at the compression-molded machine to form the composites.

Molding was done at 150 °C for 10 min. After heating, molding was immediately cooled by running cold water followed by keeping it in the refrigerator. Figure 4 represents the process of manufacturing tiles using textile waste. In the case of denim waste, it is cut into small pieces and mixed with the resin and hardener followed by the same procedure as described above. Various samples of tiles along with their codes are shown in Table 1.

2.5 *Testing of Tiles*

Following test methods were used to evaluate developed tiles:

2.5.1 Tensile Strength (ASTM D-638)

Five 10 mm width test specimens were prepared. Before the testing, specimens were conditioned at 23 ± 2 °C and 50 ± 5% relative humidity for 40 h. The width and thickness of each specimen were measured using the applicable test methods.

Fig. 4 Process of making tiles using textile waste in the laboratory

Table 1 Developed tiles sample with code

Sample Code	Sample Description
DS1	Denim waste developed tiles using synthetic resin
DN1	Denim waste developed tiles using natural resin
CS2	Cornhusk waste developed tiles using synthetic resin
CN2	Cornhusk waste developed tiles using natural resin
RS3	Rice straw waste developed tiles using synthetic resin
RN3	Denim waste developed tiles using natural resin

The test was run at 5 mm/min out to a minimum of 0.5% strain before removing the extensometers. The Instron machine used to test the strength properties of tiles is shown in Fig. 5.

$$\textbf{Tensile strength}\ (N/mm^2 \text{ or } \textbf{MPa}) = \frac{\textbf{Maximum load (N)}}{\textbf{Cross sectional area (Width} \times \textbf{thickness)}}$$

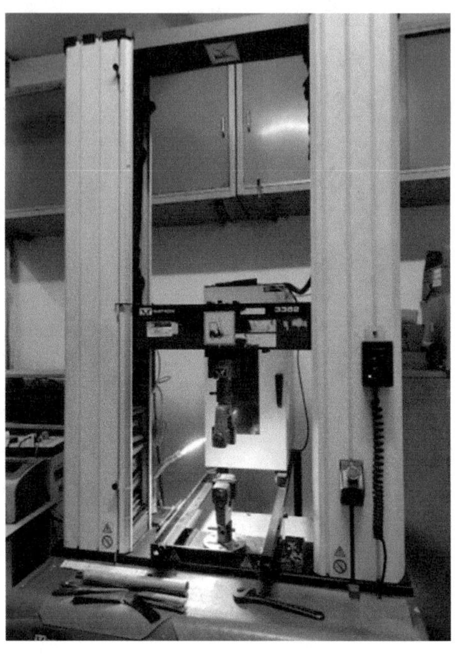

Fig. 5 Tensile testing machine

2.5.2 Flexural Strength Test (ASTM D-790)

Method 1A three-point loading system was used to test the specimen. Micrometers of least count 0.025 mm (0.001 inches) were used for measuring the width and thickness of the test specimen. Test specimens were conditioned at 23 ± 2 °C (73.4 ± 3.6 °F) and 50–65% relative humidity for 40 h before test. The gauge length and speed (5 mm/min) were set as per the specimen. Readings were noted down and calculated according to the width and length of the specimen.

$$\textbf{Flexural Strengths s} = \frac{3\textbf{PL}}{2\textbf{bd}^2}$$

where s = stress in the outer fibers at midpoint, MPa [psi], P = load at a given point on the load–deflection curve, N [lbf], L = support span, mm [inch], b = width of beam tested, mm [inch], and d = depth of beam tested, mm [inch]. Unit = N/mm2 or MPa.

2.5.3 Water Absorption Test (IS 1363 (Part 2)

Water absorption tests were carried out as per the ASTM D570 test standard. Samples of each composite type were oven-dried before its weight was recorded as the initial weight of the composites. The samples were then placed in distilled water maintained at room temperature (25 °C) for 24 h. The samples were then removed from the water, dried with cotton fabric, and weighed. The amount of water absorbed by the composites (in percentage) was calculated using Eq. (1):

$$\%\mathbf{W} = \frac{\mathbf{Wt} - \mathbf{Wo}}{\mathbf{W0}} \times 100$$

where *Wt* is the weight of the composite after water immersion and *Wo* is the weight of a dried sample.

2.5.4 Shock and Heat Stability (IS 13630 (Part 5)

Determining the resistance to thermal shock of all ceramic tiles in normal conditions of use. This is carried out as per IS 13630 (part 5) in this method, determination of resistance to thermal shock of the whole tile by cycling 10 times between the temperature of cold water and a temperature just above that of boiling water. Usually, tests are carried out between 15 ^{0}C to and 145 °C.

2.5.5 Chemical Resistance IS 13630 (Part 7)

In this method, the test specimens are partially immersed in the test solution and visual observation is done after 28 days. This test is carried out as per IS 13630 (part 7).

2.5.6 Determination of Resistance of Linear Thermal Expansion (ISO 10545-8)

The coefficient of linear thermal expansion is determined for the temperature range from ambient temperature to 100 °C.

2.5.7 Horizontal Flammability test-IS 15061, Annex A

This is based on testing of flammability behavior of the horizontally oriented sample. In this test, burnt length, burn rate, and burning time (sec) are calculated. The horizontal flame tester used in the study is shown in Fig. 6.

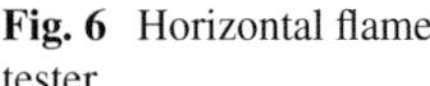

Fig. 6 Horizontal flame tester

3 Results and Discussion

3.1 Optimization of Epoxy Resin and Hardener Ratio

In this study, resin to hardener ratio was optimized. For this purpose, five resin to hardener ratios (1:1, 1:2, 1:3, 1:4, and 2:1) were applied on the 10-g weight of web of equal thickness. Out of these ratios, sample made with the 1:3 (resin: hardener) was selected as they provided optimum hardness with good surface smoothness. Below this ratio, the composite sample was delicate and flexible, and above the 1:3, the sample was becoming very hard before curing. So for further study, 1:3 ratios of resin and hardener were selected.

3.2 Development of Tiles

After the optimization of resin to hardener ratio, tiles samples were developed using a lab model composite-making machine. An appropriate amount of flame and water retardant chemicals was added to make tiles flame and water resistant. The thickness of the developed tiles was maintained as per the ceramic tiles mostly used in the buildings. The developed tiles are shown in Fig. 7.

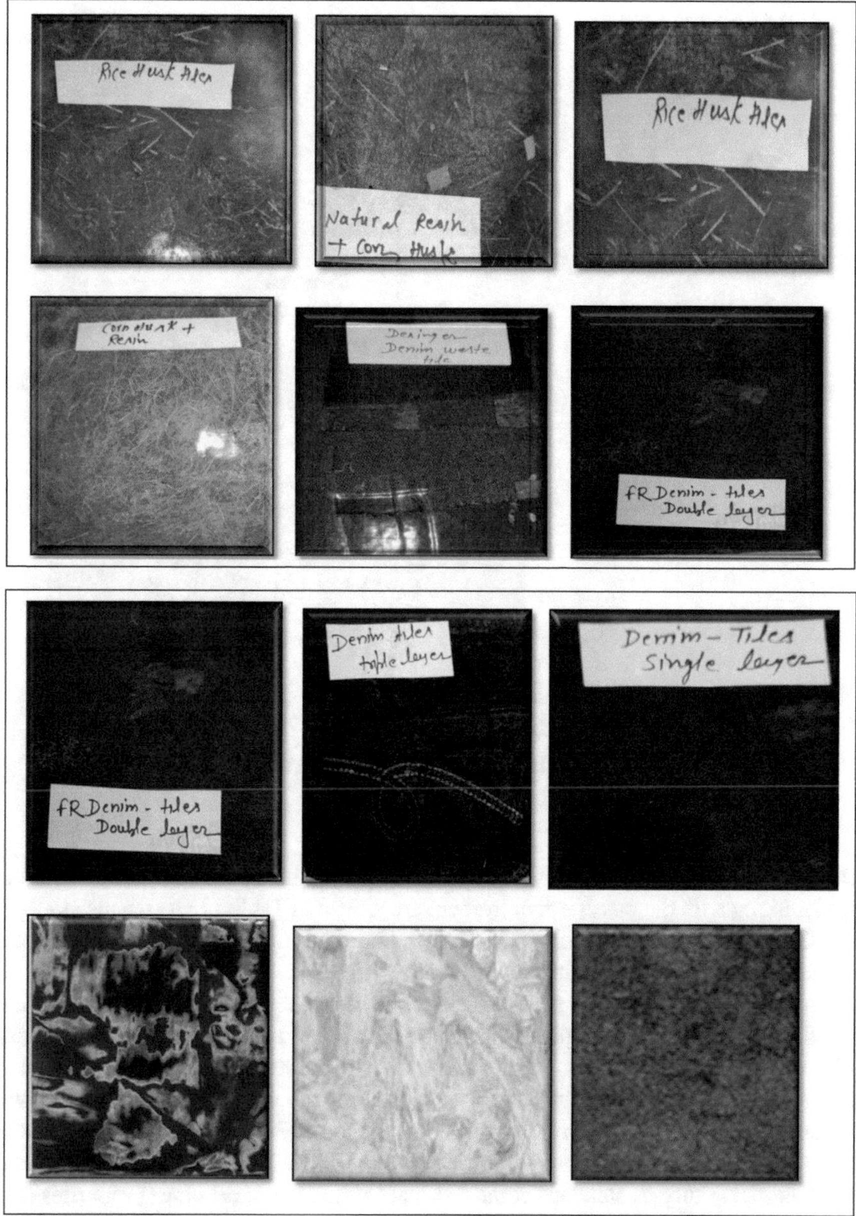

Fig. 7 Developed tiles using denim waste, rice straw, and corn husk waste

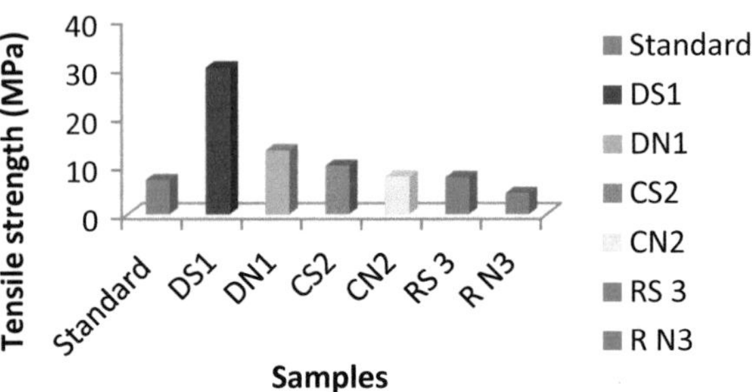

Fig. 8 Tensile strength of developed tiles samples

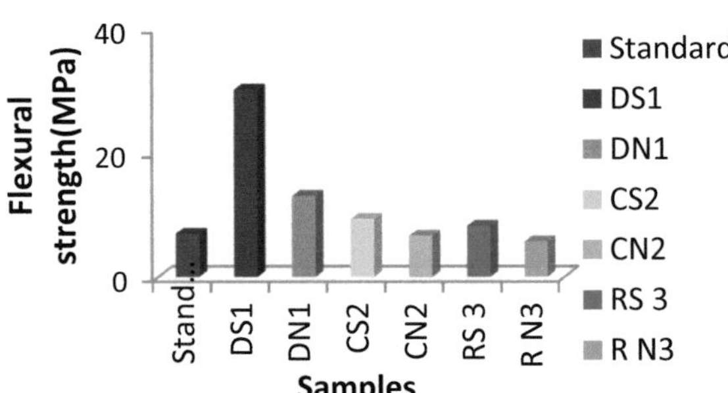

Fig. 9 Flexural strength of developed tiles samples

3.3 Mechanical Properties of Tiles Made Out of Denim Fabric Waste, Cornhusk, and Rice Straw Waste with Resin

The developed tiles evaluated the mechanical properties as per IS 13630. The test results along with the passing requirement are shown in Table 1. The tensile strength and flexural strength of corn husk rice straw and denim waste tiles with synthetic and natural resin are shown in Figs. 8 and 9.

From Table 2, it is clear that all the developed tiles meet the requirements of Indian Standards. The tensile strength of tiles made out of denim waste is 18.669 MPa, corn husk waste is 9.91 MPa, and the tiles made out of rice straw are 7.54 MPa, which is greater than the tensile strength of standard ceramic tile which is 7 MPa. It is shown in Table 2 that the flexural rigidity of denim tiles is 157.69 MPa, corn husk tiles is 56.28 MPa, and tiles made out using rice straw is 46.02 MPa, which is greater than the flexural rigidity of standard ceramic tiles. In the case of all other properties like absorbency, shock, and heat stability, the determination of chemical resistance of developed tiles meets the requirements of standard ceramic tiles.

From Table 3 it is clear that the tiles developed from textile and agro-waste using natural resin do not meet the requirement in some properties as per standard. As shown in Table 3, the tensile strength of denim waste tiles using natural resin is 9.56 MPa, corn husk is 7.66 MPa, and rice straw-based tiles is 4.33 MPa. It is clear that denim and corn husk-based tiles' tensile strength is greater than the standard

Table 2 Standard requirement as per standard test method of ceramic tiles using epoxy resin

Properties		Tiles made out of resin and textile waste			Requirement as per IS 13630 standard of ceramic tiles
		DS1	CS1	RS1	
Tensile strength	Maximum load (N)	2549.169	1275.255	970	800
	Modulus (MPa)	110	60.472	29.169	47
	Tensile strength (MPa)	18.699	9.9149	7.54	7
Flexural rigidity	Maximum load (N)	235.51	62.61	55.39	47
	Modulus (MPa)	157.69	56.28	46.02	44
	Flexural strength	30.126	9.39	8.30	6.6
Absorbency (%)		2%	2.75%	1.8%	6.2%
Shock and heat stability		Yes	Yes	Yes	Heat stable/Not shock stable
Determination of chemical resistance		Unstable to acid and alkali	Unstable to acid and alkali	Unstable to acid and alkali	Stable to acid and alkali
Determination of linear thermal expansion		< 2% of total dimensions	< 2% of total dimensions	< 2% of total dimensions	2% of total dimensions

value. But in the case of rice straw-based tile with natural resin it is lower than the standard value. These developed tiles are found to be unstable to acid and alkali resistance.

Fire Retardant and Water Repellency Test

The developed all types of tiles, made out of synthetic resin and natural resin, were tested against the fire test as per IS 15061. From the test, it was found that the flame did not propagate in the FR-treated tiles samples after the removal of an ignition source. Due to this, the maximum burnt rate was zero while in the case of tiles without FR treatments, continued to ignite after removing an ignition source. Figure 10 shows the arrangement of the FR test and Fig. 11 shows the FR testing of the sample. In the case of tiles made out of natural resin, treated with FR chemicals, all samples did not pass the criteria and continued to ignite after removing the ignition source.

All types of tiles (made out of synthetic resins and natural resin) were tested with water repellency test as per IS 390. All are found to be having a rating of 100, which means there is water repellent in nature.

Table 3 Standard requirement as per standard test method of ceramic tiles using Natural resin

Properties		Tiles made out of resin and textile waste			Requirement as per IS 13630 standard of ceramic tiles
		DN2	CN2	RN2	
Tensile strength	Maximum load (N)	1303.28	985.68	557.01	800
	Tensile strength (MPa)	9.56	7.66	4.33	7
Flexural rigidity	Maximum load (N)	87.266	44.53	37.86	47
	Flexural strength	13.09	6.68	5.68	6.6
Absorbency (%)		2%	2.75%	1.8%	6.2%
Shock and heat stability		Yes	Yes	Yes	Heat stable/Not shock stable
Determination of chemical resistance		Unstable to acid and alkali	Unstable to acid and alkali	Unstable to acid and alkali	Stable to acid and alkali
Determination of linear thermal expansion		< 2% of total dimensions	< 2% of total dimensions	< 2% of total dimensions	2% of total dimensions

4 Conclusions

Textile waste and agricultural waste generated by extracting unconventional fibers can be used for making floor and roof tiles. Three types of materials such as denim fabric waste, cornhusk fibers waste, and rice straw fibers waste were used for making tiles using synthetic and natural resin. These tiles were made fire and water resistant. The mechanical and chemical properties of these tiles were evaluated using IS 13630 specification. The results indicated that developed tiles (synthetic resin with textile and agro-waste) meet the requirement of the Indian Standard. According to tensile test results, it was found that tiles made by denim waste using natural or synthetic resin in both cases have a higher value of 18.89 MPa as compared with the tiles made by corn husk (9.91 MPa) and rice straw (7.54 MPa). Denim waste and agro-waste tiles made out of natural resins did not meet the requirement of standard in some properties such as chemical resistance and heat and shock stability.

Fig. 10 Arrangement of FR test

Fig. 11 Testing of FR test

References

1. Yalcin-Enis et al. (2019) Risks and management of textile waste: the impact of embedded multinational enterprises. In: Gothandam et al. KM (eds) Nanoscience and biotechnology for environmental applications, environmental chemistry for a sustainable world. Springer Nature, Switzerland AG
2. Parmar MS, Shukla S, Sisodia N, Jain A, Singh M (2020) Manufacture of blended yarns from waste biomass of corn. Colourage 62:34–38
3. Song HJ et al (2009) Biodegradable and compostable alternatives to conventional plastics. Philos Trans R Soc Lond B Biol Sci 364(1526):2127–2139
4. Aishwariya S, Jaisri J (2020) Harmful effects of textile waste. Fibre2Fashion, June
5. Pichardo PP, Martínez-Barrera G, Martínez-López M, Ureña-Núñez F, Ávila-Córdoba LI (2017) Waste and recycled textiles as reinforcements of building materials, natural and artificial fiber-reinforced composites as renewable sources, Dec 20
6. Ms. Hetal Mistry (2020), Textile value chain, upcycling of pre-consumer textiles waste, July 7
7. Sachdeva PK, Chanana B, Parmar MS (2014) Extraction of textile grade cellulosic fibres from sugarcane bagasse, International conference on emerging trends in traditional & technical textiles, technical textiles, NIT, Jalandhar (India), April 11–12
8. Chang Y, Chen H, Francis S (1999) Market application for recycled post consumer fabrics. J Res Fam Con Sci 27:320–337
9. Nemerow NL (1963) Theories and practices of industrial waste treatment. Addison-Wesley, Reading, MA
10. Mohammed L, Ansari MNM, Pua G, Jawaid M, Saiful Islam M (2015) A review on natural fiber reinforced polymer composite and its applications. Int J Polym Sci 2015:1–15
11. Gadkari R, Burji MC (2014) Textiles waste recycling, textile value chain, June 6
12. Palakurthi M (2016) Development of composites from waste PET - cotton textiles. Textiles, merchandising and fashion design: Dissertations, Theses, & Student Research, Dec
13. Todor MP et al. (2019) Recycling of textile wastes into textile composites based on natural fibres: the valorisation potential, Mater Sci Eng 477:012004
14. Wan Y (2010) Fiber and textile waste utilization. Waste Biomass Valor 1:135–143
15. Thacker H (2019) Up cycling Denim waste. Indian Text J, Mar 29
16. Meng X (2019) Recycling of denim fabric wastes into high-performance composites using the needle-punching nonwoven fabrication route. Text Res J 90(5–6):695–709
17. Peña-Pichardo P et al. (2018) Recovery of cotton fibers from waste Blue-Jeans and its use in polyester concrete. Constr Build Mater 177:409–416

Sustainable Development Goal: Sustainable Management and Use of Natural Resources in Textile and Apparel Industry

Shanthi Radhakrishnan

Abstract The United Nations has envisaged a sustainable development plan for the year 2030 which initiates 17 sustainability development goals (SDGs) with objectives that promote all round development. This forum encourages contributions from all sectors—governments, industrial, civil organizations, public and private sectors—as opportunities for the fulfillment of these goals. The textile and fashion industries have been very popular in the extensive use of natural resources accompanied by waste and waste products that tend to pollute the environment causing hazards to the living organisms in the planet. Businesses and brands in the textile and apparel sector are earnestly working on aligning their production and management on the basis of sustainability, the pinnacle being the sustainability development goals. This chapter deals with the sustainable management and effective use of natural resources (SDG 12—Target 12.2)—water, energy and soil for the development of sustainable textile fibers and certification methodologies for sustainable reporting (SDG 12—Target 12.6). This can be achieved by sound management of chemicals and wastes occurring in the production cycle or life cycle of a product (SDG 12—Target 12.4). Green productivity in sustainable manufacturing calls for improved resource efficiency and waste reductions by implementing a cleaner manufacturing strategy. The specialized long value chain of the textile and fashion industry is poised to address the sustainability challenge to achieve the economic, social and environment development goals.

1 Introduction

In the year 2000, the millennium development goal was initiated on a global scale, with the aim to deal with the humiliation of poverty. The combined efforts of the world helped in accomplishing development and eradicating poverty among the lives of many people, but the program remained unfinished. Efforts to continue the good work of the MDGs were essential giving rise to the post-2015 development agenda

S. Radhakrishnan (✉)
Costume and Apparel Designing, PSGR Krishnammal College for Women, Coimbatore, India

S. S. Muthu (ed.), *Sustainable Approaches in Textiles and Fashion*, Sustainable Textiles: Production, Processing, Manufacturing & Chemistry,
https://doi.org/10.1007/978-981-19-0530-8_8

which has integrated the core essence of sustainability—sustainability development goals (SDGs) [1]. In 2012, the UN Conference on Sustainable Development held in Rio de Janeiro formed the starting point of the Sustainable Development Goals. In September 2015, the UN General Assembly adopted the sustainability development goals—a 15 year agenda till 2030 [2]. With 17 goals and 169 targets, the SDGs are broader in scope and tend to address and eradicate the root causes of poverty leading to development suited to all people around the globe. The new goals are universal and encompass the three dimensions of sustainability—economic development, social inclusion and environmental safety. Powerful focus on implementation is a key factor with the SDGs in terms of resource mobilization, competence building and expertise complemented with institutional organization for data analysis and reporting frame work. The impetus and success of the MDGs have led to the formulation and implementation of the SDGs.

During the SDG Summit in September 2019, world leaders felt that though progress was made in many places, the achievements were not progressing at the required speed or scale to reach the targets by 2030. A Decade of Action was prescribed on three levels [3], namely **global action** for better leadership, additional resources and smart solutions for the SDGs; **local action** which calls for the necessary changes and conversion of policies, financial plans, regulatory and institutional frameworks of governments and local authorities; and **people action** at the grassroots level which includes all stake holders to facilitate and extend their support and enthusiasm to achieve the acceleration for fulfillment of the goals. The Sustainability Development Report showcases the global assessment of the progress of the member countries toward achieving the SDGs and further strengthened by the SDG indicators and the voluntary reviews provided by the nations [4]. The Sustainability Development Report 2021, the seventh edition, unfolds with four ***Chapters*** [5] (Part 1—Increasing the Fiscal Space of Developing countries to achieve the SDGs; Part 2—The SDG Index & Dashboards; Part 3—Policy efforts and Monitoring Frameworks for the SDG; Part 4—Methods Summary & Data Tables); ***Rankings with interactive maps*** [6, 7] of all the 193 UN Member States assessed as overall score and spill overs. The overall score shows the total progress of a country toward achieving all the 17 SDGs. The spill over score shows the positive/negative effects of a country's actions on the other countries' ability to achieve the SDGS. This is assessed over 3 dimensions as impacts in terms of environmental, social and security. A higher score shows more positivity; ***Country Profiles*** [8] track the progress and trends in achieving the SDGs in the member nations; ***Data Explorer*** [9] provides the norms set for each SDG and shows the nations under the each category along with their achievement status from 2010 to 2020. The reports, supplementary materials, SDG indices and databases are all downloadable to readers.

Ensure sustainable consumption and production pattern, SDG 12 [10, 11], promotes efficient management of resources and energy, sustainable infrastructure, access to basic services, green jobs and better quality of life. This goal aims at all round development and reduction of poverty. Currently, awareness of natural resource consumption has increased and efforts have begun to curtail air, water and soil pollution. 'Doing more and better with less' is the focus of SDG 12 and this calls for higher net gains in economic activities when care is taken to reduce use of resources, degradation and pollution along the entire life cycle, thereby bringing in a change in the quality of life. This requires help to the producers to the final consumers through education on sustainable consumption and lifestyle change, provision of information through standards and labels and making it easy for sustainable public procurement.

SDG 12 has 11 targets and 13 indicators as specified by the UN. The targets are the goals and the indicators are the metrics by which the goals are achieved. The targets and the metrics are given in Table 1 [12].

The online data base for the Sustainability Development Report 2021 [13] reveals that the extent of goal achievement for SDG 12 among the 206 entries was 27 (13.23%) denoted by the color green; yellow color denotes 'challenges remain' being 51 (25%); orange for 'significant challenges' being 55 (26.96%); red for 'major challenges' was 53 (25.98%); and gray showing no information being 18 (8.8%). The aim is to slowly remove challenges and move all participating nations into the green color which signifies 'goal achievement'. The indicators used for assessment were municipal solid waste (kg/capita/day), electronic waste (kg/capita), production-based SO2 emissions (kg/capita), SO2 emissions embodied in imports (kg/capita), production-based nitrogen emissions (kg/capita), nitrogen emissions embodied in imports (kg/capita) and non-recycled municipal solid waste (kg/capita/day).

Raw material conservation from natural sources is of primary importance as it results in safeguarding the environment. According to the SDG Report 2020, the Global Material Footprint has increased from 73.2 billion metric tons in 2010 to 85.9 billion metric tons in 2016 (+17.3%) [14]. Biodiversity loss—over 31,000 species are in the extinction list which is 27% more than the species indicated in the IUCN Red list (1–6). Agricultural expansion has caused the decline of forest areas at an alarming level; between 2015 and 2020, each year 10 million hectares of forest areas have been destroyed [15]. The report also mentions that the fossil fuel subsidies have risen from $318 billion to $427 billion being one of the primary contributors to climate change crisis. The COVID-19 crisis has offered opportunities to concentrate on recovery plans to build a sustainable future. This principle has been activated as the report also states that between 2015 and 2017, 79 countries and the European Union have contributed to *sustainable consumption and production* with change in at least one policy [14].

Environmental crimes are destruction of the environment and squandering of natural resources by illegal activities. To quote a few examples [16]:

- The illegal activities in the Amazon Rain forest include clearance of land, illegal logging, wild life mining and trafficking which affect biodiversity and climate change. These activities are accompanied with extensive violence, corruption, human slavery and money laundering.
- In western Guyana, the village of Etheringbang is notable for illegal gold mining conducted under the watch of 'Sindicators', the Columbian ELN and Venezuelan mafias, along with the Venezuelan security forces. All supplies pass through check points where extortion payments are charged at gunpoint.

Table 1 SDG 12 Targets and indicators [12]

Target	Target description	Indicator	Indicator description
12.1	Implement the 10 year framework of programs on SC&PP with all countries taking action by 2030	12.1.1	Number of countries that have a SC&P national action plan
12.2	Achieve sustainable management and efficient use of recourses	12.2.1	Material footprint per capita and per GDP (sum total of material footprint for biomass, fossil fuels, metal and non-metal ores)
12.3	Halve global per capita food waste	12.3.1	Global food loss index (reduce food loss at retail and consumer level; post-harvest losses; along production and supply chains
12.4	Achieve environmentally sound management of chemicals and all waste through international frameworks	12.4.1	Percentage of countries meeting the commitments and obligations within each agreements
		12.4.2	Hazardous waste generated per capita, proportion of hazardous waste treated, type of treatment
12.5	Sustainably reduce waste generation through prevention, reduction, recycling and reuse	12.5.1	National recycling rates, tons of material recycled
12.6	Encourage companies to adopt sustainable practices and reports	12.6.1	Number of companies publishing sustainability reports
12.7	Promote sustainable public procurement practices	12.7.1	Number of companies implementing public procurement practices and action plans
12.8	Promote universal understanding of sustainable lifestyles in harmony with nature	12.8.1	Extent to which sustainable education is mainstreamed (global citizenship and sustainable development education)
12.A	Support the developing countries for scientific and technological capacity for SC&P	12.A.1	Amount of support on research and development for SC&P
12.B	Develop and implement to monitor sustainable development impacts for sustainable tourism	12.B.1	Number of sustainable tourism strategies/policies/action plans with the prescribed evaluation tools
12.C	Rationalize inefficient fossil fuel subsidies that encourage wasteful consumption	12.C.1	Amount of fossil fuel subsidies per GDP in proportion to the total national expenditure on fossil fuels

- The state of Roraima, in Brazil, has no legal mine but has exported 771 kg of gold during the last 3 years. There are about 5000 illegal miners and illegal mining has resulted in elevated mercury poisoning of over 90% of the villagers tested, increase in homicide rates and displacement along the Brazil-Venezuela border.

- Brazil's Amazon has witnessed illegal seizures of both public land and indigenous areas and clearing of forest for cattle pasture. The number of cattle in the Amazon has reached 86 million in 2018 which is 4 times more when compared to 1988.

Illegal mining is one of the main drivers which has accounted for one quarter of the deforestation in the Amazon. The violence against indigenous people increases by about 150% in 2019, while deforestation grew by 25% in the beginning of 2020. It has been reported that about 2,540 violent incidents have been documented targeting environmental defenders which included 406 assassinations [16]. The destruction in the Amazon region has reached the pinnacle and the urgent need of the hour is to undertake efforts to interrupt and dismantle environmental crimes.

It may be noted that SDGs have brought about transformations in many business markets. H&M has brought about a series of changes and targets, namely use of 100% sustainable cotton (organic, recycled or certified by bCI) by 2020; raising the target for collection of used clothes to 25,000 tons per year by 2020; promotion of innovation, e.g., re:newcell; focus on energy efficiency, renewable energy and carbon sinks for absorption of greenhouse gas emissions; fair living wages; vegan leather; foundation 500; recycling blend textiles by hydrothermal process; and so on. These efforts have resulted in the titles 'One of the world's most Ethical Companies' by the Ethisphere Institute, for the seventh time in a row, Freedom House Award for leadership and global supply chain transparency.

The H&M group's sustainability work has been noted by those outside the company. In 2017, the H&M group was named as one of the world's most ethical companies by the Ethisphere Institute—for the seventh year in a row. The group also won the Freedom House Corporate Award for its leadership in advancing global supply chain transparency. The American organization Freedom House gives the award to recognize businesses for their principled policies and strong leadership in the area of human rights. The H&M group was ranked third in the Fashion Transparency Index, which evaluates supply chain transparency among the world's 100 largest fashion companies and also in terms of environmental and social impacts. H&M is also listed in the Dow Jones Sustainability Index—World & Europe, FTSE4Good and also in the Corporate Knights—Global 100 Most Sustainable Corporations in the World [17].

The above is an example, but there are many institutions and companies which are striving hard and relentlessly to fulfill the SDGs targets and work toward sustainability.

2 Natural Resources—Use and Management

2.1 Use of Natural Resources

Any material from the planet Earth that is used to meet the needs and support humans can be considered as a *natural resource*. Natural resources are used to make food (plants and animals), fuel (coal, natural gas and oil) and raw materials (metals, stone, wood and sand) for the making products. Some of the results of extracting and using virgin resources from nature may cause pollution of land, water or air, disturbance/extermination of ecosystems and a reduction in biodiversity, which deals with the variety of plant and animal life in the world. The Sustainable Europe Research Institute (SERI) report, 2009, states that the annual natural resource extraction was 60 billion tons of raw material, which is around 50% more than what was extracted 30 years ago; further, the prediction is that the extraction of natural resources could increase to 100 billion tons by 2030. The report also highlights the consumption behavior of rich versus poor countries of the world. Rich countries consume 10 times more natural resources than the poor countries, e.g., North America (90kg resources/day) and Europe (45 kg resources/day), while Africa consumes 10 kg resources per day [18]. Case studies also reveal the negative environmental and social impacts of extraction in developing and emerging countries, namely oil extraction in Nigeria, copper mining in Peru and palm oil production in Indonesia and Malaysia. The challenge facing the world is to ensure a high quality of life for the current global population and for the increasing population of the future, without surpassing the environmental resource capacity of the planet.

Let's see what happens when we overuse resources. The most common greenhouse gas is carbon dioxide emitted while burning coal, oil and natural gas. The solar heat is absorbed and retained by the greenhouse gases leading to increase in the global temperature of the atmosphere near the Earth's surface. This condition over time can cause serious dangers like flooding, drought and disease. The current commitments will lead to around 3°C global warming by the end of the century which would be fatal and catastrophic to small islands and coastal areas. The climate change drama has unfolded by way of Caribbean storms, wild fires and droughts in Africa [19].

Overuse and extraction of natural resources can ring alarm bells within ecosystems. Any biological community with organisms that live and interact with each other in the same physical environment is termed as an ecosystem. While studying nature, the ecosystem is the most basic and fundamental unit. It is made up of two components, namely the *biotope or abiotic constituent,* which deals with a physical environment with specific physical characteristics like climate, temperature, pH, nutrient composition, humidity, etc., and the *biocenosis/biotic constituent* consisting of a set of living organisms like animals, plants, microorganisms, which are continuously interacting and interdependent with each other [20]. This relationship can extend to any scale like any living organisms/plants with lakes and water bodies; mountains and forests with Mother Earth. Preservation of ecosystems is of primary importance as humans are dependent on the ecosystems for survival. For cereals, vegetables and agriculture to take place, natural processes like pollination, soil characteristics, temperature and humidity are important. If these conditions are drastically changed, then the results may become harsh and irreversible. Since all the living organisms are dependent on the environment and ecosystem, a balance has to be maintained. Techniques like agroforestry, permaculture and regenerative agriculture are built on these concepts to maintain and build a robust ecosystem.

Ecosystems provide water, food (cattle and sea food), products (pharma, biochemicals and industrial products) and energy (sunlight, biomass and hydropower) as in Fig. 1. The most important regulating services offered by ecosystems are climate regulation where oceans, trees and soil facilitate the absorption of carbon and storage; the microbial process occurring in the soil–waste decomposition; contribution by bees and other insects to help crop pollination and reproduction; regulatory processes for water and air purification and control of pests and diseases. Supporting and habitat services include primary reproduction, nutrient and seed dispersal. Cultural services that touch the inner consciousness of human beings are inspiration for creativity, entertainment and spirituality; ambience for introspection, silence and reflection; feeling of goodness and happiness; recreation, adventure and ecotourism; scientific innovation based on the principles of nature-biomimicry.

To explain the importance of the ecosystem and its preservation, we have the story of how humans had affected the ecosystem of the US Yellowstone National Park. As a predator control measure, the US Biological Survey killed the wolves and other species resulting in the disappearance of the wolf population which lead to a number of chain reactions—Tropic Cascade. It is an ecological phenomenon which starts at the top of the food chain and reaches down to the bottom. When the wolves vanished, the deer population grew tremendously grazing the vegetation to nothing. After 70 years, the wolves were introduced into the park in 1995. The wolves not only kill animals but also give life to many other species. They killed some of the deer and radically changed the behavior of the deer. The deer started avoiding the valleys and the gorges, the areas where they could be spotted and hunted easily; once the grazing diminished, these places started to regenerate—the height of trees became 5 times more in 6 years and the barren lands gave rise to willow, cottonwood and aspen trees beckoning the birds and the beavers which feed on trees. The beavers created opportunities for other species like otters, muskrats, fish, reptiles and amphibians. The wolves killed coyotes increasing the number of mice and rabbits leading to the

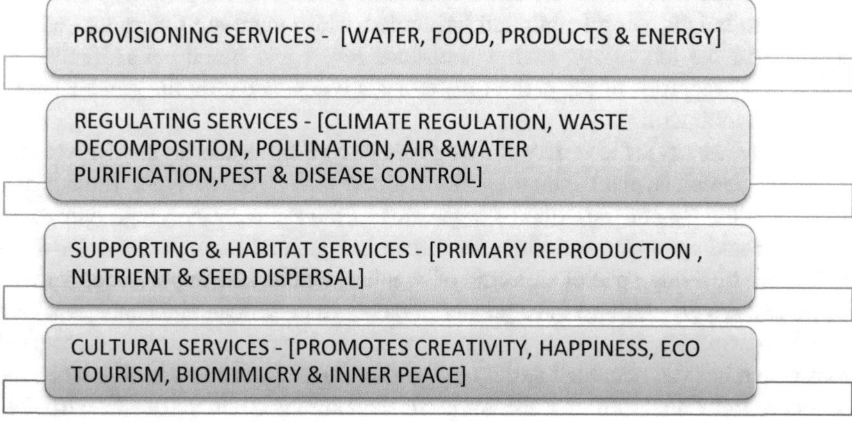

Fig. 1 Services offered by ecosystems

arrival of more eagles, foxes, hawks, badgers and weasels. The ravens, bald eagles and bears fed on the carcass left by the wolves. The berries from the shrubs and the abundant food led to the population increase of the bears. Further, there was a change noticed in the behavior of the rivers. Lesser erosion caused lesser meandering and narrowing of channels forming numerous pools and rifle sections, creating a haven for wildlife habitat. The regenerative forests on the stabilized banks helped in marking the courses of the rivers; the vegetation in the valley and gorges flourished resulting in lesser soil erosion. On the whole, the wolves transformed the ecosystem of the Yellowstone National Park and its physical geography. The rivers changed in response to the wolves [21], the efficient ecosystem engineers. There are similar stories where human activities have caused severe disruption to the ecosystems.

Diminishing the biodiversity is another result of greedy extraction of the natural resources for industrialization. Biodiversity is built on three main concepts, namely ecosystem, species diversity and genetic diversity. When these three parameters are strong and interactive, the biodiversity will be resilient. The components of a biodiverse system ensure the health of the ecosystem and its inhabitants by processes like purification of water and air, control of diseases and pests, support pollination and enhance fertility of soil and storage of carbon. Biodiversity is the core which makes a place habitable, healthy and liveable. Currently, ecosystems are encroached and altered by human activity changing and impacting the phase, characteristics, specificities and functions of an environment.

Genetic diversity is the variety of genes that exist among living organisms. There are differences in genes, and their expression is also different among species and within different species. The variance in genes gives rise to a wide range of life forms and variation in physical and biological characteristics based on the interaction with the environment. The term 'phenotype' is the observable physical characteristics of an organism like appearance, development and behavior. Phenotype is determined by the type of genes and the environment which influences their development [22]. Even though the genes are identical, e.g., identical twins, they may express dissimilar phenotypes based on the environmental influence. Some of the phenotypic characteristics are height, wing length and hair color, while there are others that can be measured in the laboratory, such as hormone levels and blood cells. Hence, the quality and properties of the natural resources are dependent on the genetic set up and the environment.

Species diversity is the variety of living species on Earth and is divided into groups like insects, animals, plants, fungi.... The diversity may be of two types: intraspecies diversity is the genetic variation of organisms of the same species, e.g., hair color, skin color and eye color variations in human beings, and interspecies diversity is the variety of living species in terms of number, nature and importance; namely, human species (7.7 billion) have greater diversity than African elephants (reaching extinction). Species richness is the number of species in an ecosystem, e.g., tropical areas which have an abundant variety of species. Species evenness is high when the number of individuals within a species is constant among communities. The tropical rain forests contain almost half of the world species with 5–10 million insect species, 40% of world's flowering plants, 30% of the world's total bird species; the Great

Barrier Reef of Australia houses 400 coral species, 1500 fish species, 4000 species of mollusks and 6 turtle species. It covers 0.1% of the ocean but has 8% of the world's fish species highlighting species richness [23]. More diverse and rich the ecosystem, it can produce more for use, helps in distributing the resources maintaining equilibrium and can withstand environmental stresses leading to stability and sustainability.

Varieties of ecosystems contribute to ecosystem diversity. It takes into account the number and nature of the ecosystems. The different ecosystems of the Earth include oceans, deserts, lakes, forests and plains, and within each type, there are different environment specificities like hot/cold deserts, Mediterranean/ tropical forests and so on. Each ecosystem has its own specifics, functions, species and peculiarities. When biodiversity is assessed, the parameters taken are all types of living organisms and their interaction and contribution to the ecosystem.

The Amazon Rain forest, one of the most biodiverse regions on Earth, has complex ecosystem, huge mix of species and the genetic variety within the species. It has been estimated that the Amazon Rain forest produces 20% of the total oxygen in the atmosphere of the Earth through photosynthesis [23]. In this forest, there are huge vines which start from the trunk of the trees to the canopy, intertwining with the treetops and their thick wooden stems supporting the trees and branches; the trees provide the fruit and seeds which are carried by the herbivorous animals like Tapir and Agouti, to different parts of the forest enhancing the growth of new plants and providing food for millions of insects [24]. The rain forest is a huge system built on smaller systems packed with interconnected species; every link provides stability to the next system strengthening biodiversity weave. This weave is further reinforced by the genetic diversity within individual species, which helps them to cope with changes. However, there are vulnerable species that lack genetic diversity due to isolation or low population, which are unable to withstand fluctuations due to climate change, habitat fragmentation or disease. In some ecosystems, removing one keystone species will cause the links to fall apart and unravel the entire ecosystem. For example, the coral reefs nurture many organisms by providing key ingredients, microhabitat, nutrient, shelter, breeding ground for thousands of fish, mollusks and cry stations; corals are interdependent with fungi and bacteria enabling a sturdy weave that supports various organisms. Thus, the coral, a keystone organism, if weakened or killed by destructive fishing, ocean acidification or pollution, can in turn affect its dependents and the entire ecosystem. Thus, the ecosystem, genetic and species diversity together form the complex weave of biodiversity vital for the survival of organisms on Earth. Humans are also woven in this weave as they are dependent on nature for resources. Extraction and abundant use of resources without concern for regeneration can disturb the relationships within the ecosystem and facilitate biodiversity loss. Biodiversity helps in ensuring Earth's own protection net to safeguard the survival of all organisms including humans.

In 1972, the UN Conference on the Human Environment held in Stockholm, Sweden, realized the need to regulate the use of natural resources in order to preserve ecosystems and its components. Some of the salient features of the Stockholm Declaration [25] are well represented in Principles 2, 3 and 5. Principle 2 states that air, water, land and the ecological network (fauna and flora) and any other as applicable, the natural resources of the Earth, are to be protected for current and future generations by appropriate careful planning and management. Principle 3 indicates that the capacity of the Earth to produce essential renewable resources (flow resources like water, air, natural energy forms and biomass) should be maintained/restored/improved for optimum use. Principle 5 refers to non-renewable resources of the Earth (coal, petroleum, natural gas, etc.) must be used judiciously to prevent complete exhaustion and to be shared by all mankind. The Stockholm Declaration has addressed many issues like resource depletion, sharing of resources within and across countries, intergenerational equity and a balance between sustainable use to maximize social benefits and minimize environmental impacts.

2.2 *Natural Resource Management*

We have seen the story behind natural resources and how they work, let us see how we can work toward sustainable resource management. In order to work toward sustainability, we have to understand the term 'natural resource regeneration'. Any natural resource needs time to regenerate/revive/renew for further use, e.g., cutting branches of trees and estimating the time taken for it to come back to its form before cutting. Hence, it is essential to understand that all renewable resources require time to regenerate. If the renewable resources are not given time to regenerate and is again and again used, it leads to 'natural resource exploitation'. Hence, there needs to be a balance between the demands of exploitation and regenerative capacities. It is hard to set limits to this issue because of the fact that the Earth has limited resources, its regenerative capabilities and the absolute diversity of natural resources that make scientific assessment difficult.

Basically, natural resource management has two main concepts, namely maximum sustainable yield (MSY) and optimum utilization (OU) [26]. All natural resources have an optimum utilization level or acceptable levels of use. This level is usually set by management authorities or national/international regulatory authority based on scientific knowledge with a purpose of regulating exploitation. The MSY is a regulatory principle that introduces real-life data into the management process. Let's take the example of fisheries. When death of fish occurs due to human harvesting, the reproduction rates will rise. This is the peak level of production after which the overall population will decrease culminating the positive effect of increased rate of production. When the surplus production is harvested sustainably according to the MSY that is specific to each population, there will be good regeneration of resources that will stay for a long, long time. If the population decreases due to problems like disease or habitat loss, the harvesting should be done according to the current MSY. Hence, careful monitoring of population growth and overall health of the population are required. Measures like reducing the harvest of females or immature fish can ensure the population to be maintained to the required sustainable levels.

Use of natural resources may be consumptive or non-consumptive. In the first case, the resource is consumed effectively that it cannot be put to any other use. Here, management would be balancing exploitation in relation to regeneration capacity of resources. In the second case, humans do not consume the resource, but may interact with them. Management will be in the form of regulation of human interaction and control the negative impacts of such interactions on the resource. Here, management trend will be protection versus exploitation. Whales were used as a raw material and fuel in the West till the mid of the twentieth century; today, they are a recreational resource in the West and as food in other parts of the world. The uses of resources may change and require change in management and outlook. Resources that are localized like forests and collective resources that do not have specified borders namely air & sea, need extra special managerial solutions; selfish motives of excessive usage of these resources can lead to environmental degradation, pollution, and climate change which impact on a global scale. Here, limiting the exploitation of resources to make them available for future is not possible as individuals have the tendency of using large quantities of collective resources before the others get a hand to it. A blend of private and communal management is required which may give rise to problems in resource management.

3 Sustainable Agricultural Practices for Reduced Land Eco-Toxicity

Agricultural practices are considered to be sustainable when the farming practices prevent negative effects to soil, water, biodiversity, other allied resources and the impact on living organisms. This may include regenerative farming, organic farming, permaculture, agroforestry, biotechnology in agricultural practices, mixed farming, multiple cropping and crop rotation. Let's discuss regenerative farming, biotechnology in agriculture and organic farming.

3.1 Regenerative Farming

Regenerative farming is a combination of farming and grazing practices that reconstruct and build the organic matter of the soil, thereby boosting its biota diversity. This enhances the capacity of the soil to absorb, store and sequester atmospheric carbon and increase its water holding capacity. Further, there is a two-way benefit, namely the reduction of CO2 content in the atmosphere which is the cause of global warming and the improvement of soil structure that has been damaged by human civilization. The farming practices can be measured for carbon enhancement of the soil, and the raw material produced in this land is termed as 'Climate Beneficial' [27]. Carbon Farm Funds help in the implementation of farm practices that produce climate beneficial materials, e.g., North Carolina Fibershed. Farm and ranch scale contributions fall between 1 and 3% of the annual point of sale purchase while mid- and large-scale donations include price premiums of 60 cents per pound. Many farmers adopting regenerative farming practices have been given the title of 'carbon farmers'—Lani & Estill of Bare Ranch, California [28]. Reports indicate that climate beneficial production has a carbon footprint differential of 150 pounds of CO2 per garment when compared with the conventional methods of production [27, 29].

Some of the regenerative practices applicable to fiber farms are compost application on pasture, silvopasture, rotational grazing, conservation tillage, pasture cropping, planting windbreaks and buffers and creek restoration. A half inch layer of compost is applied to the pasture, which slowly releases the fertilizer causing an annual carbon sequestration rate of one ton per hectare for 30 years without reapplication [30]. Silvopasture is a manmade combination of pasture and trees for sequestering large amounts of carbon in the soil; managed rotational grazing allows animals to graze in planned pasture divisions for longer rest periods. Value creation in the minds of the consumer is enhanced by the term 'climate cooling wool' produced by Fibershed for the manufacture of climate beneficial products. Taking the cue, the North Face has introduced its Cali Wool Collection—beanie, scarves and jackets using 'Climate-Beneficial TM Wool' from Fibershed. Regeneration International, a nonprofit organization in the US, networks with 4 million members in more than 100 countries, provides hands-on and online training in sustainable farming and runs a data bank of American and European sustainable fashion brands [29] for eco-conscious customers.

3.1.1 Factors Affecting Regenerative Agricultural Practices:

Water

Fresh water is a scarce resource and requires careful usage. Soil becomes salinized and infertile due to continuous irrigation techniques. Desalination is a technique used for producing fresh water for agriculture. It has been reported that in 2015, 3–5kWhm^{-3} energy was used to produce 100million m3 of fresh water serving as the feasible alternative to irrigation [31]. In the desert and dry areas, water deposited during the night and early morning evaporates during the day. This can be preserved by covering the surface with a layer of fiber waste which cools more quickly, and combined with the porous nature of sand, the absorption of water is higher; the surface of the fiber waste tends to provide space for water to condense. It has been reported that 13% came from the land-surface source and 19% from dew in the semi-arid areas, while conditions of fog can condense more water. A manmade layer of algae in the Qubqi Desert fetched two times more rainwater when compared to the conventional setup [32–35].

Soil Organic Matter

Sequestration of carbon in the soil is an important factor for better farmlands. The soil organic matter (SOM) includes the organic components, namely undecayed plant and animal tissues and their partial decomposition products along with the soil biomass [36]. A study in the arid grasslands reported that the SOM above the ground (155.3), below the ground (95.3) and total was 256.3gm^{-2}; the mean soil organic carbon density was 1.38 Kgm^{-2} in the top 30 cm of the soil. During cultivation, organic waste is added to increase SOM resulting in increased carbon from the organic waste and carbon sequestration from the vegetation. The estimated sequestration of carbon would last for 155 years. The addition of cover crops would have a potential global soil organic carbon sequestration, which could compensate 8% of the direct annual GHGs from agriculture. This sustainable system is beneficial for grasslands and also decreases waste masses. The waste mass serves to increase the moisture content helping the microorganisms to thrive well and create an environment for the cultivation of plants and trees in the semi-arid areas [37–41].

Cultivation Methods

Cultivation methods play a very important role in the successful harvest of any crop. Two methods have been discussed as examples for sustainability. In the first method called Zia, planting is carried out in huge pits (dia 50 cm and 45 cm depth), which are dug during the dry season to act as a carrier for water harvesting and organic matter. The organic matter is filled in the pit and further up to give a cone-like appearance on the tree trunk/ base of the bush planted in the rainy season. This process is self-reinforcing and helps in SOC process year by year. The second method is to spread a 0.2-m-thick layer on the surface of the area, which promotes water absorption and increases SOC [39, 42–44].

No till or low till farming is a sustainable agricultural technique that leaves the top soil undisturbed for the growth of mycorrhizal fungi which helps the symbiosis of crop roots and earthworms to provide better outcomes. The farmer tills about one feet deep and loses the crop residue by 90% and also leads to erosion by wind and water.

Carbon Content of Textile Fibers

Textile waste materials are used as a resource for energy. The amount of carbon released into the atmosphere while burning textile fibers is dependent on the cellulose content of the material; for example, cotton consists of 90% cellulose [44]. One ton of cotton when burned releases (1000kg $\times$ 90% $\times$ 0.444) 400 kg t^{-1} carbon and 1.47 t CO_2. Synthetic polymers and wool tend to produce more amount of CO_2. It has been estimated that the textile incineration (municipal solid and textile waste) is around 78650t leading to CO_2 gas of 115600t per annum [45, 46]. The cotton waste

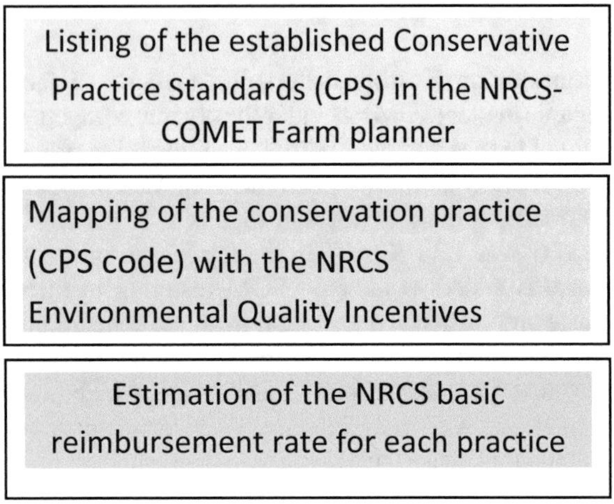

Fig. 2 Funding system for conservative practice standards [50]

can be converted into fertilizer to improve the SOM. The density of compressed textile waste is estimated to be 265kgm^{-3}, 17,700 kg cotton can supplement the biomass on 11.4 ha at the cost of \$270 to \$600 per hectare, and the breakeven cost as a cheap alternative will be achieved by 20 years. When the fertility of the arid soil is increased by such agricultural practices, the farmlands fetch a better price.

Uses of Textile Waste

The Waste Framework Directive (WFD) recommends the following system to be followed with regard to waste segregation, namely prevention, reuse, recycling, other recovery and disposal. In Sweden, the textiles left in the municipal solid waste were reported to be 77100 tons out of a total consumption of 121000 tons. A majority of this waste was burned in power plants by incineration. This could be used for improving the SOM. Technological advances have helped in segregating waste textiles according to the inherent raw materials. Near-infrared reflectance spectroscopy (NIR) has been used to identify components of textile waste. In 2017, Fibersort is running a 5000 t capacity facility for waste segregation. NIR also helps in estimating the carbon content of the waste collected [45–49].

Implementation of Conservation Practices—Costs and Funding

Regenerative farming practices for wool production in North Carolina are an interesting aspect for analysis. When a conservation practice is introduced, the ranchers generally want to know about the benefits of adopting the practice and what will be additional value that will be incurred. A basic rate for the conservation practice was fixed as mentioned in Fig. 2. A farmer who has a 10 acre holding, non-organic and 80 sheep may be interested in two conservation practices like 'cover crops – CPS 340' and 'mulching-CPS 484' for which the basic reimbursement rate is $61.91 and $188.28 per acre, respectively. The farmer will use 4 acres ($274.64) for cover crops and 2 acres (376.54) for mulching leading to a total of $624.20. The non-reimbursement funds are usually around 50–85% ($312.10–530.57), which may be earned by the higher price of wool or from grant funds. This encourages the farmer to start the conversion and slowly convert the non-organic farm to a sustainable one.

A regenerated fiber fund was set up by Fibershed to fund on-ranch conservation practices in three different models. The first model is *Regenerative Premium on Raw wool,* which pays an additional cost over the market value ($0.60/pound) to the rancher who produces regenerative wool. The rancher benefits from the immediate returns, but it add costs to the supply chain increasing the price at different stages (up to $2.40/pound). This model is considered as most effective. The second model is *Regenerative Wool Premium on Textile Products* adding value to textile programs. Here, 3% is added to the textile sales price which is paid to Healthy Soils Fund initiated by Fibershed to distribute to the ranch partners. The funds help Fibershed to build regenerative fiber arrangements and also create awareness by consumer education through a point of sale model. This system was taken up by Houston Textiles, which eventually became a tax deductible donation for the company. This model was less effective in raising money for the ranchers but helped in creating awareness among the consumers and general public. This system facilitated Fibershed by creating funds for future regenerative wool programs. The third model is a donation to the Regenerative Fiber Fund either as an $8 tree donation per linear foot of tree planted area or a $50 'compost donation' for one cubic yard of compost application [50]. This model includes a wide range of supporters who are not into the fiber processing business but are interested in promoting regenerative products. This model is considered least effective but helps in raising capital to fund future practices.

3.2 *Biotechnology in Agriculture*

Biotechnology is a valuable tool for the implementation of sustainable agriculture. The production systems have been drastically changed with the advent of biotechnological advancements. The biotechnological shift helps to nurture the environment, enhance the quality of life and also instill economic progress. In 2005, about 21 countries had grown transgenic crops like maize, canola, cotton and soya bean of which 14 countries had covered more than 50,000 hectares with these varieties. In 2005–2006, about 28% of the total global agricultural area was growing biotech cotton varieties (insect resistant IR and herbicide resistant HR), which represented 37% of the total cotton bale production and 38% of the global bales exported [51, 52]. The main aim of developing IR and HR cotton was to produce plants what could control lepidopteron insects and to protect the plant from a variety of herbicides using gene technology. Cotton varieties are available with multiple output traits like agronomic performance, gossypol reduction, fiber quality and abiotic stress tolerance.

Agronomic sustainability is based on rotation of crops, integrated pest management IPM, use of multiple technologies and cultural practices leading to the prevention of over farming and soil erosion which are the perennial threats to the cotton-cultivated areas across the globe.

The major goals for sustainable cotton production are environment positive, economic and viable and enhancement of the quality of life, which can be achieved by catering to the growing demand of the fiber, maintaining the natural resource base and adopting sustainable farming operations. Cotton faces the challenge of being replaced by manmade synthetic fibers if it does not maintain the qualities of production and operations coupled with biotechnological developments. Reports state that the use of IR cotton varieties shows that there was a 14% (−1.2 million kg) reduction in the use of insecticides in six US states showing improvements in surface and ground water quality which would otherwise be contaminated by runoff and leaching of residual chemicals [53, 54]. This feature is shown in Table 2 and Fig. 3. The use of HR cotton has increased the yields showing an increase in pounds and a reduction in the herbicide application. While working on the economics, five studies taken in seven states have shown that on an average the increase in returns was $8.42 /hectare with an increase in yield by 9%. The introduction of IR cotton proved to be beneficial as the profit earned was around $99 million in the year 1999 when compared to the conventional cotton production and cost in the previous year [55, 56]. The benefit of using IR cotton with stacked genes could be understood by the report which states that the $28.70 was the economic advantage or $4.23/hectare [57, 58]. A similar

Table 2 Cotton herbicide application [1994–1999] [59]

Year	Lbs/acre	Applications (per acre)
1994	2.02	2.96
1995	1.89	2.91
1996	1.57	2.53
1997	1.7	2.84
1998	1.61	2.76
1999	1.63	2.82

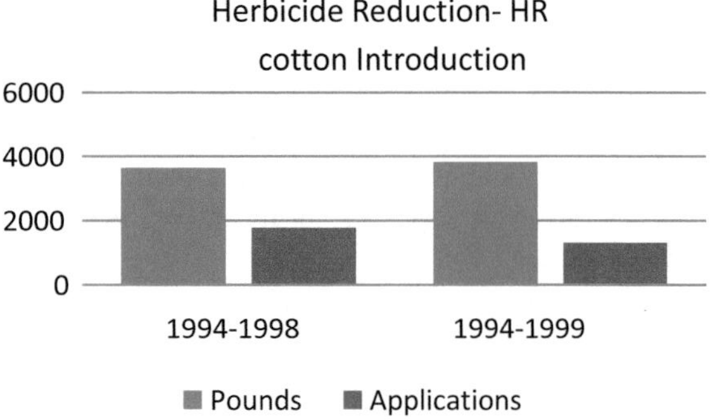

Fig. 3 Carpenter and Gianesse [53]

sequence was seen in 2005 when India expanded the plantation of IR cotton hybrids by 160% with the cultivable area of 1.3 million hectares by one million farmers. ICAR conducted many studies, and the cost–benefit analysis of IR cotton showed an increase of 67% in gross income on an average with yield increases from 62 to 92% [59].

The quality of life is a social aspect in the drive to sustainability. BT cotton has brought safety to human health in terms of reduced pesticide use and exposure. In China, the incidence of symptoms by pesticide poisoning significantly reduced by around 80% demonstrating the health advantage of using IR cotton [60, 61]. Further gossypol content in cotton seed makes it unsafe for animal and human consumption [62]. It has been estimated that the annual yield of 44 million metric tons of cotton seed with 9.4 million MT of protein could address the total protein requirement of half a million people @ 50 gm/day. Biotechnology is helping to bring about this conversion for human health and nutrition and to cater to the food requirement of the world population which has been predicted to increase by 50% in the next 50 years. Another important feature is the ease with which IR/HR cotton can adapt to technology and also holds reduction in labor inputs, e.g., it takes 3 h to stack a bale when compared to 25 h in conventional cotton [63].

3.3 *Organic Farming*

A twentieth-century development is organic farming which relies on the concept of nature-based agriculture. This system uses growth boosters of organic origin and also uses sustainable methods of agriculture. Some of the organic fertilizers used are compost manure (manure from animal feces), green manure (manure from plant origin and farming left outs) and bone meal (ground animal bones and slaughter house remains), and the techniques adopted are crop rotation (growing different crops in the same area), companion planting (planting different crops in proximity), biological pest control (controlling pests with natural predators), use of natural insect killers and elemental synthetic substances (copper sulfate, sulfur, lvermectin). Substances like genetically modified organisms (organisms with modified genes), nano-materials (substances between 1 and − 100 nm), hormones, human sewage sludge (semi-solid remains after any process), plant growth regulators (signal molecules in low concentration which promote growth) and use of antibiotics in animal husbandry (synthetic medicines for livestock) are prohibited in organic farming [64–66]. Organic agriculture is internationally monitored with a set of standards by International Federation of Organic Agriculture Movements (IFOAM) founded in 1972 [67].

Organic farming works for soil fertilization and health; enhancement of biological activity by the use of cover crops, green and animal manure and crop rotations; weed management and insect and disease control by the use of biological means and crop rotations; better augmentation of biodiversity of system with environment; improved livestock operations and health care of animals using rotational grazing and mixed forage pastures; reduction of waste and 'of the farm inputs' through elimination of chemical fertilizers, pesticides, synthetic medicines and antibiotics. The whole perspective in organic farming is an emphasis on renewable resources, conservation of soil and water, and agricultural practices that enhance environmental equilibrium. There is a growing awareness among all sectors of the population about the benefits of organic farming and products, and it has been reported that in 2016 about 57,800,000 hectares have been farmed organically, which represents 1.2% of the total land around the globe [66, 67].

Organic cotton initiatives were started in Asia, Africa and S. America due to market demand and retailers option for organic garments. In 2006, the production of certified organic cotton was 31,000 tons and the turnover for organic cotton garments reached US $583 million [68, 69]. Trials were conducted to check the productivity difference between organic cotton and conventional cotton. The report given by the Central Institute of Cotton Research, Nagpur, and a pilot study in Andhra Pradesh, India, showed higher yields 11-21% and 13%, respectively, in the case of organic cotton [70–72]. In India, the gross margins were higher for organic cotton when compared to conventional cotton by 52% in 2003 and 63% in 2004 reducing the financial risk for the farmer. Organic cotton also had a premium price and the intermittent crop also contributed to the profit. However, there are obstacles in the initial stages of organic cotton cultivation. The conversion from conventional farming to organic farming takes 3 years when the yields may decline from 10 to 50% [73, 74]. This is usually attributed to the time necessary for the soil fertility and ecological balance to respond to organic nutrients and pest management; the farmer also needs training in organic farming and management and requires hard work, immense patience to bear with the pressure of weed and insect management during the initial stages.

4 Water and Energy Management

4.1 Water as a Natural Resource

Sources of water that are useful to humans are considered as ‘water resources’. We see so much of water around us that it gives an impression that it will last very long, say a million years. The Earth is referred to the ‘Blue Planet’ as three fourth of the surface is made up of water [75]. Data indicates that fresh water contributes to only 2.5% of the total water on Earth and two-thirds of this water is found in glaciers and at the polar regions. Fresh water may be from two sources such as surface water (ponds, rivers, lakes and streams) or ground water (water that seeps into the ground and trapped between soil and rocks). It has been estimated that 70% of the water in the world is used for agricultural irrigation [76]. The impact of climate change will be heavily felt on the water resources as there is a close connectivity between climate and the hydrogen cycle. Use of water by mankind has depleted the aquifers and many pollutants are a great menace to water supplies.

Since water is a renewable resource, it is comforting to think that we can get back water if used with great care. Limitless exploitation of this resource can lead to depletion. The United Nations Development Plan states that when annual water supplies reach below 1700m^3 per person, it is termed as ‘water stress’, and when the annual supplies drop below 1000m^3 per person, we can say that the area is experiencing 'water scarcity'. The prediction is that there will be severe water scarcity in Africa and West Asia by 2025 due to high water demands [77]. Another school of thought predicts that the planet Earth could become barren and dry like the planet Mars within one billion years as the oceans are slowly going into the interior of the planet. The Tokyo Institute of Technology has estimated that annually 1.12 billion tons of water moves from the oceans into a layer of rock found in the interior of the Earth called the Mantle [78]. This movement has caused the sea levels to fall down by 600 m in the last 750 million years. Scientists also believe that water can return back to the surface through volcanos and mid-ocean ridges, and annual return of 0.23 billion tons, which is far below the required level. Water being precious should be used with utmost care and managed well to prevent wastage.

4.2 Water Management

The raw material for fashion and apparel industries is textiles, and cotton is primarily used for a majority of apparels. World Wildlife Fund (WWF) has indicated that 73% of the cotton supply of the world is nurtured in plantations using natural irrigation. This will in turn reduce the supply of water over a period of time. The Aral Sea in Central Asia [79, 80] is the best example where the waters of the feeding rivers Amu Darya and Syr Darya was diverted to grow cotton and agriculture. The Aral Sea, the fourth largest inland water body, dried up creating an ecological catastrophe. The sea shrunk by 1994 to form two entities and the water and vegetation became salty; the animals who ate the grass fell ill and many animals would bang their heads on the ground and die. After some time, the Kokaral Dam was built between the Large and Small Aral seas that caused water to flow back to the Small Aral Sea. Lesser salt, fish came back and people who left the place started coming back. However, the impact has been great and the people are waiting for some miracle to occur—to see water in the Aral Sea to make it once again the fourth largest inland water body of Central Asia.

Steps to improve carbon footprint and reduce water wastage:

- Agricultural innovations: use of recycled water for cotton plantations, save water through new technologies.
- Raw material: use of environment friendly raw material, e.g., biocotton; unconventional sturdy fibers, e.g., hemp, linen, nettle and flax that require less water fertilizers and pesticides; recycled fibers can cut the water footprint when compared to virgin fibers. Osomtex, a US-based yarn brand, has zero water in the process as it follows mechanical recycling. Similarly, EcoAlf, Spain, has marched forward by recycling plastics from the ocean into clothing [81, 82]; upcycled fibers to produce less water intensive fabrics. In 2016, Levi Strauss & Co. teamed up with Evrnu for producing Levi's 511 prototype from five discarded cotton t-shirts. The cotton waste is pulped and then extruded into a new fiber which is finer in denier than silk and stronger than cotton [83] and 98% less water usage compared to

virgin cotton products. Blue Ben, German label has opted for a biodegradable viscose developed by a Portuguese firm Tintex, thereby saving 94% water that is consumed by conventional cotton [84].

- Use of new technologies in textile manufacture. Wet air oxidation is a viable method that converts pollutants into water and CO2 at high temperatures of 300 oC and high pressures of over 10 MPa for efficient conversion within the stipulated time without additional sludge or concentrated waste as by-products [85]. Can be used for treatment of all wet processing wastewater of the textile industry. Similarly, SpinDye, Sweden, has chosen to dye polyester before spinning leading to a reduction of C02 emissions by 30% and water consumption by 75% [86].
- Sourcing of sustainably grown cotton based on international standards, e.g., fair trade certified or Better Cotton Initiative and so on.
- Partnering with manufacturers who recycle and reuse process water, keep processes on no water or least water menu [87]
- Technological improvements and research should be directed to find ways of recycling water and reuse in day-to-day manufacturing. Favorable and commercially sustainable practices and their success should be made aware to the textile and apparel fraternity for similar initiatives and attempts [88].
- Eliminating hazardous chemicals in clothing manufacture by incorporating Detox best practices in chemical, water or textile-related regulation (national or global legislation) with rewards for sustainable followers.
- Chemical and sourcing transparency and safe alternatives: adoption of hazard elimination roadmap, research for safe alternatives and closed-loop production processes.
- Increasing the lifespan of individual clothing through reselling and rental. Efforts have been taken by Selfridges, London, to become a zero emission business enterprise by 2050. The retailer has vowed that by 2025, it will have products that are made from materials that adhere to sustainability standards and will be assessed continuously as per the requirements specified by certification bodies and green groups [89]. Selfridges has partnered with Vestiaire Collective, second-hand market place for luxury fashion in 2019 and in 2020 launched 'Reselfridges' where customers can buy and sell luxury second-hand articles. Further, in May 2021, Selfridges has associated with Hurr, a rental platform, to start 'Selfridges Rental' which enables customers to rent in the season stock. Similar reseller market place has come into existence like Levi's Secondhand, COS Resell, etc. A report states that if the life cycle of cotton garments is extended by nine months, the water footprint of clothing can be reduced by 5–10% [90].
- All manufacturers, retailers and suppliers need to estimate their products and services in terms of sustainability impacts—environmental, social and economy. H&M, in June 2021 has followed the HIGG Index ranking for their products similar to the traffic light system which will provide an insight to customers regarding the water use and pollution impacts of their products. A scoring system is used; the baseline score indicates that the materials that constitute in making the product are not biodegradable, while products with scores 1,2,3 show that the biodegradable materials have been used in the product, with 3 showing the highest

degree of sustainability. The report states that six products in the H&M website have score 3.

In the COP24 climate conference in Poland, 43 companies signed the UN Fashion Climate Charter with the aim of achieving 'net zero greenhouse gas emissions' by 2050. Water scarcity is a looming threat and it has been predicted that by 2025 two-thirds of the world's population will have to live in water-stressed regions [20]. Greenpeace's focus on water pollution and its reduction followed by UN's declaration naming the decade as 'Water Action Decade' have increased awareness and responsible action.

The UN Global Compact has announced the CEO Water Mandate, with an aim to mobilize business leaders on the issues of water, sanitation and SDGs. As on date, 203 companies have endorsed to work continuously toward six core elements of water stewardship. This will enable all members to understand to manage one's own water-related threats. The six core areas of water stewardship are: (1) direct operations—include those who directly or indirectly use water for the production of goods and services; (2) supply chain and watershed management—include water-related activities along the supply chain; (3) collective action—encourages collaboration between stake holders across sectors and societies; (4) public policy—provides inputs and develop policies at local, national and international levels; (5) community engagement—initiates community and local-level awareness and education on water; (6) transparency—reporting, accountability and sharing of water strategies [91]. Each core area has a description of the goal, target and pledge that the members have to endorse for continuous effort and action.

The Water Action Hub is an online global knowledge sharing platform with 1694 projects, 1065 organizations at 5595 locations in different parts of the world [92]. The hub helps in promoting awareness of organizations and their sustainability projects, facilitation of new projects, partners and collective action to address water-related risks and sustainable management of water and join the Water Resilience Coalition to reduce water stress. Users/members can access the projects from any part of the world and understand the problems, process and technical solutions to face the challenges of water stewardship. Thus, water as a natural resource requires serious deliberation before use, and its management is very critical for regeneration as a resource for future use. Many organizations and individuals are facing the impact of overuse of water and are looking forward to stringent measures to protect and save this precious resource for future applications.

4.3 Natural Resource: Energy

The next important natural resource is energy which is the most needed item in today's world. Matter that stores energy is named as fuel. Energy is the capacity to do work and can be in many forms, namely thermal and radiant obtained directly from the sun, electrical and mechanical acquired indirectly from the sun, chemical and atomic energy which are independent and not from the sun. Atomic or nuclear energy is the heat generated by the splitting of atoms (fission) or by the fusion of two atoms. The sun's energy is created by the fusion of atoms. The by-product of nuclear fission is radioactive waste which will last for 500,000 years. Nuclear fission has been existing from World War II [93, 94].

Natural resources from the energy perspective are termed as primary fuel resources—coal, oil, natural gas and uranium. Their occurrence on the Earth is in the form of deposits that have to be discovered and extracted. These fossil fuels are non-renewable and predicted to decline. It has been estimated that 90% of world's energy needs are fulfilled by fossil fuels; they provide 66% of the world's electrical power and 95% of the world's energy requirement leading to an alarming statistics of doubling consumption of fossil fuels every twenty years [95, 96]. The primary energy flow resources are the wind, hydro and the solar and are freely available as there is no reservoir unlike the primary fuel resources [97]. Solar energy can be used in many forms for different applications, namely thermal heat, ventilation (stack effect), shading (blocking the direct sun), radiant cooling (absence of sun in outer space at night), day light (natural light), photo chemical (reactions of sun and materials) and humidification (evaporation using energy from the sun). A tiny fraction of one percent of solar energy, which is absorbed by the Earth, is changed into plant tissue that has produced all the fossil fuels. Harvesting fossil fuels and burning them for energy can be hazardous to the environment. It will take thousands of years for the Earth to reproduce the consumed fossil fuels as they are non-renewable resources [93].

Air mass is another natural resource for the production of energy. The radiant heat from the sun is unequal on the surface of the Earth; this produces differences in the heat contents and density of the air mass causing movement that is amended by various factors like rotation of the Earth, inclination, distribution of land and oceans followed by geographic abnormalities. This movement of the air mass is termed as wind. Wind energy can be used in many ways like cooling, promoting combustion, as windmills for pumping water and electricity. Likewise, the movement of water in its natural cycle can create energy, namely hydro-electric dams, turbines, ocean thermal gradients, tidal power, thermal inertia and thermal storage by water.

Economic development and energy consumption go hand in hand. Generally, economy grows to meet global competitiveness and is based on availability of low cost, environmentally safe energy sources and the ability to meet energy demands. The need for renewable energy is progressing at a rapid pace, and the focus is on enhancing the distribution and transmission infrastructure and improvement of conventional, hydro and atomic energy followed by energy storage. Whatever may be the energy demand, the production rate of a resource must be studied to understand the replenishment time. This will keep the resource alive and supply the resource consistently. Hubbert's curve [98] is a model that estimates the production rate of a resource over a period of time. In this estimation, the Hubbert's peak refers to a point where the production rate will be the highest with the demand for the resource rising after which there will be a drop in production compared to the demand. This was originally devised for the US oil production. Hence, more consideration should be evident to control the use of resources in an optimal way to enable the replenishment of the natural resource.

4.4 Energy Management

Industry and the nation will obtain substantial benefits if the efficiency of energy is increased. To understand and assess opportunities for increasing energy efficiency, one must identify how the industry is using energy, amount used for various systems, loss percentage and quantity that goes into processes. Next, the energy that can be recovered through energy efficiency, technology and other means is to be assessed. Research is being conducted by the US Department of Energy's Industrial Technologies Program (DOE/ITP) to fasten environmentally viable and energy efficient technology for industry through a multiphase study to identify the usage of energy sources and systems, target opportunities for reducing the use of energy and increase the conversion of energy resources into useful production output. This can be best understood in Fig. 4.

To start a study on energy recovery, primary energy and fuels and electricity have to be taken into account. Primary energy represents all the processing energy related to the industrial plant boundary, both internal and external. Primary energy use comprises of power and fuels and the losses that occur at the site from where the

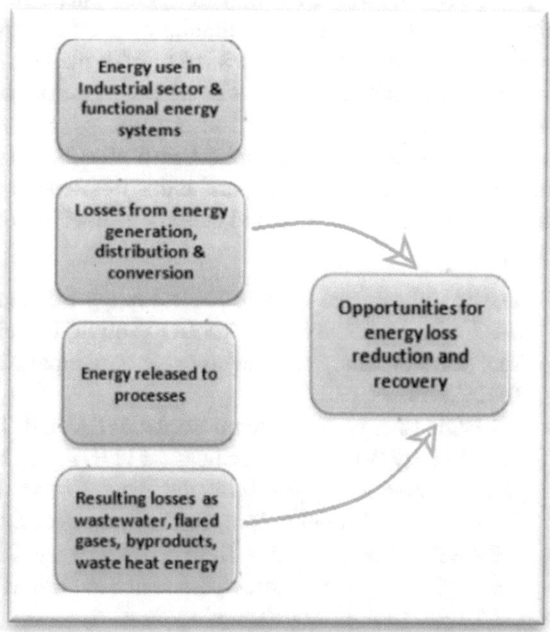

Fig. 4 Energy use, loss and opportunities in industrial production

energy is purchased. However, it does not include the energy used as raw material often termed as 'feedstock energy'. The next part of the study is the comparison which identifies industries as either electricity intensive or fuel intensive. The potential for onsite cogeneration technology can be understood. The onsite energy losses usually occur in process heating and cooling and hence one must take into account steam systems, fired systems and the cooling systems. The potential downstream losses are usually not given due care and can be as much or more than those prior to the process. Opportunity analysis will help to reduce and recover the energy losses to a great extent to bring about cost reduction and savings.

Coming to the textile industry/mill, conversion of electrical energy into mechanical energy is important. Usually, the conversion efficiency is 85% and this should be stepped up to provide monetary savings. The type of motors and their output play a significant role in energy savings. Energy conservation can be achieved by minor change of equipment, high capacity utilization and technology upgradation. In India, two forms of energy are used, namely electricity (spinning) and fuels (wet processing), and the share of energy expenditure in manufacturing cost is increasing and has moved from 12% in 2003 to 18–20% in 2017 [99]. Let's take the case of the US textile industry steam and motor-driven systems account to 28% of the total energy use and the estimation of onsite energy losses is around 13% in motor-driven systems, 8% in distribution and 7% in boilers [100].

While considering energy efficiency, the utilization of light source, use of multiple motors with control board and electric heating vs other methods should be examined in the case of motor-driven systems; for efficient fuel use, selection of fuel and boiler to prevent pollution is efficient; with regard to steam losses, selection of transportation pipes, installation of steam accumulators and heat exchanger are important. Use of non-conventional energy sources is a must as maintenance cost will be low, pollution will be almost nil and there will be no wastage. An energy audit will help to identify problems and promote energy conservation.

Use of energy is always linked with greenhouse gas emissions. Managers and engineers in the textile and apparel industries need to track their energy usage and in turn the GHG emissions. To aid these personnel, a suite of tools is available, namely [101]

- Textile and Apparel-specific Tools: Energy Efficiency Assessment & Greenhouse Gas Emission Reduction (EAGER) Tool; SET Tool; Energy Distribution Support Tool (EDST); Energy Management & Benchmark Tool(EMBT); Self-Assessment Tool (SAT).
- US DOE Energy Assessment Tools: MEASUR Tool; 50,001 Ready Navigator Tool; Energy Performance Indicator LITE (EnPI LITE) Tool; 50,001 Ready Navigator; Energy Footprint Tool; the Plant Energy Profiler Excel (PEPEx) Tool; Steam System Modeler Tool; the Process Healing Assessment and Survey Tool (PHAST); MotorMaster + Tool; AIRMaster + Tool; Pump System Assessment Tool (PSAT); Fan System Assessment Tool (FSAT).
- USA EPA Tool: Energy Tracking Tool.

Thus, the water footprint and energy footprint help in bringing to the notice of the industry many salient ways and technologies to conserve the natural resources. These resources are scarce and subject to high price volatility. However, the most appropriate and effective technologies are not implemented due to limited knowledge and awareness about how to incorporate efficiency measures into the day-to-day functioning of the industry. Expertise related to water and energy efficiency technologies and practices, implementation techniques and their results have to be disseminated to all concerned to motivate and encourage them to follow and attain benefits.

5 Certification of Sustainable Textile Fibers

Many certification and standards have been set up for sustainable textile fibers of which a few will be discussed.

5.1 Textile Exchange

Textile Exchange is a nonprofit organization initiated in 2002 with headquarters in Texas and has branches in 11 countries and 25 member countries. Some of the standards of sustainable concern are Content Claim Standard CCS, Organic Content Standard OCS, Responsible Down Standard and Responsible Wool Standard. The content claim standard verifies the organic integrity of input materials as mentioned by the vendor. The organic content standard verifies the accurate amount of organic substance in the final product. The CCS and OCS go hand in hand with each other. The Responsible Down Standard RDS ensures that the birds have been nurtured well to express innate behaviors, healthy living with no pain, fear or distress. This standard follows the product from the baseline to ensure that the raw material comes from ducks and geese that are reared in a responsible manner. Similarly, the Responsible Wool Standard (RWS) identifies the best practices of the farmers and checks the wool comes from farms and sheep treated in a responsible way. Textile exchange also recognizes the SDGs and incorporates the requirements as per the regulations [102, 103].

5.2 The Organic Trade Association (OTA)

is an association that promotes and protects organic agriculture with a vision to achieve excellence in agriculture and commerce and protect the environment for the well-being of the society as a whole. With its headquarters in the US, it represents 9500 organic business across 50 states. Organic 101 (eco label) is a standard that has given specific federal requirements for growing and processing organic products [104, 105]. The National Organic Standard Board NOSB states the allowed and restricted substance list for verification and understanding. A detailed application needs to be given by the organic business organization after rigorous announced and unannounced audits by inspectors and consumers. The system is transparent and checks from the agricultural aspect in terms of growing, post-harvest, preparation, processing and handling of the product in par with the federal organic standards. Annual reviews and inspection follow after issuing the organic certificate.

The National Organic Program NOP covers the production of agricultural commodities like cotton and wool in the raw material stage but does not include regulations for manufacturing stage. Hence, this cannot be applicable to garment manufacturing and textile processing. Manufacturers look out for certification from GOTS which has standards and regulations for all stages of apparel or product development (spinning, knitting, weaving, dyeing and finishing) with organic fiber. Both these two standards are taken in union to make it 100% organic.

5.3 International Federation of Organic Agriculture Movements (IFOAM)

The federation, founded in 1972 with headquarters at Bonn, Germany, has affiliates in over 120 countries. The primary aim is to work toward sustainability in agriculture right from the field to the consumer, building awareness and helping the transition of farmers to organic agriculture. The administrative structure of the of the federation includes a World Board which appoints affiliates and gives directions to undertake work—regional bodies to formulate alliances on a regional scale and sector platforms for specific priorities.

IFOAM standards include the IFOAM family of standards which are for global use and for countrywide use. The global standards are International Standard for Forest Garden Products (FGP) and Biocyclic Vegan Standard. There are standards which are country specific as given in Table 3

The Family of Standards is officially endorsed by the Organic Movement and Standards with both private standards and government regulations as required. The IFOAM Standard and IFOAM Accreditation have services for certifiers [106, 107].

5.4 Global Organic Textile Standard

This standard (GOTS) was introduced in the year 2006 encouraged by the growth of organic fibers and the demand for a wholesome certification suitable for industry and retailers. The organization is comprised of 4 reputed organizations: Organic Trade Association (OTA), USA; Internationalen Verband der Naturtextilwirtschaft

Table 3 Family of standards—worldwide [106]

Sl. No	Name of the Standard	Country
1	Tunisia Organic Regulation East African Organic Product Standard The SAOSO Standard, South Africa Zimbabwe Standard for Organic Farming, Zimbabwe	Africa
2	Asian Regional Organic Standard Saudi Arabia Organic Standard China Organic Standard India Organic Standard Israel Organic Standard Japan Organic Standard HKDRC Organic Standard, Hong Kong Biocert International Standards, India ACT Basic Standard, Thailand	Asia
3	National Standard for Organic and Bio-Dynamic Produce, Australia New Zealand Organic Export Regulation Australian Certified Organic Standard, Australia NASAA Organic Standard, Australia Pacific Organic Standard	Oceania
4	EU Organic Regulation Switzerland Organic Regulation Turkey Organic Regulation Bio-Suisse Standard, Switzerland Krav Standards, Sweden	Europe
5	Argentina Organic Regulation Costa Rica Organic Regulation Ecuador Organic Regulation Canada Organic Regulation USA Organic Regulation DIA Organic Standards, Argentina IBD Organic Guidelines, Brazil	The Americas

e.V. (IVN), Germany; Soil Association, UK; and Japan Overseas Cooperative Association JOCA, and other international stake holders with rich expertise in organic farming [108, 109]. Universal recognition for this system was due to the ease of getting one certification for organic textiles and products bearing an assurance of reliable quality and consumer acceptance. Any textile product with 70% organic content can be certified by this system and caters to the environment and social side one. Environmental and toxicological criteria govern the ecological side of the system with reference to input material like dyestuffs and auxiliaries and output like residues and waste water. The system monitors with a ecolabel, the growth and harvest of raw materials till the product reaches to consumer. All stages of manufacture in the textile supply chain are examined based on the quality assurance, revision principles and review procedures.

5.5 Organic Farming Certification

In India, this certification is governed by the National Program for Organic Production (NPOP), Ministry of Commerce and Industry. The organization provides standards for organic production, systems, criteria and accreditation. The standards have been formulated in harmony with the international standards and those complying the procedures can use the organic logo for their products. The organic crop production will include the crop production plan, conversion requirements, duration of conversion period, landscape, choice of crop and varieties, diversification in crop production and management plan, nutrition management, pest, weed and insect control, contamination, soil and water conservation and collection of non-cultivated material of plant origin. Based on the inputs, assessment and review, the certification will be provided with a logo.

5.6 The Sustainability Framework of Food and Agriculture Systems (SAFA)

The FAO has developed a universal framework to assess food and agricultural systems SAFA. After 5 years since inception, the organization has provided the SAFA guidelines, indicators and tools and app that are user friendly and easy to participate. Since this framework is adaptable, the Cacique Guaymallen water management program in Argentina used this tool to assess the sustainability of the local hydrological system working in union with FAO regional office for Latin America and the Caribbean [110]. The area was divided into sustainability polygons, and different trials were undertaken. SAFA framework was used to select the most sustainable alternative for decision making and adoption. The sustainability goals fall under the umbrella of FAO.

5.7 Field to Market: The Alliance for Sustainable Agriculture

Field to Market: The Alliance for Sustainable Agriculture is a platform for creating a sustainable agricultural supply chain for continuous increase in productivity, environmental concern and societal well-being. The food, farming and agriculture industry will benefit and create safe and eco-friendly food, fiber and fuel. The main fundamentals of this organization are supply chain transparency, worldwide collaborations and technological tools and resources that works toward continuous improvement. The trend in this organization is to bench mark, catalyzing and enabling sustainability claims.

The assessment framework called Fieldprint Platform (Fig. 5) helps retailers, suppliers and brands to check at every stage and the environmental impacts of the crop

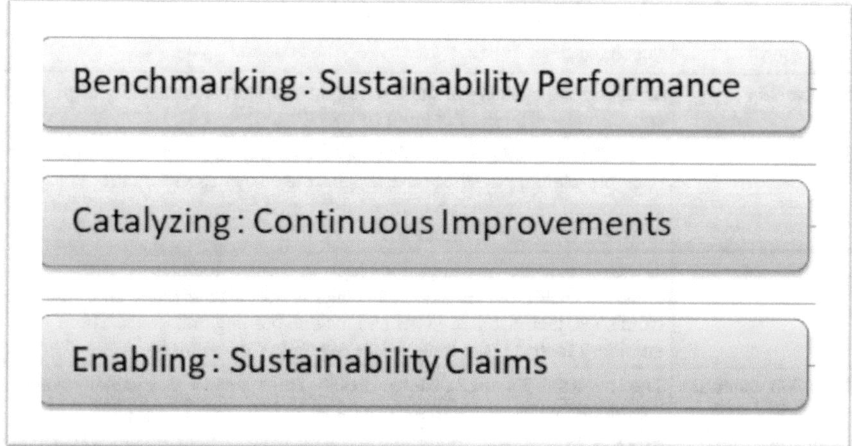

Fig. 5 Goals of Fieldprint Program [111]

production for continuous improvement. Eight sustainability metrics are embedded in this platform as given in Table 4.

These metrics are service-based tools that measure the environmental outcomes of individual farms and have been in use for over a decade. The Metrics Committee reviews the metric once in three years and revisions, and updations are made based on field experience. The Metrics Standard Operating Procedure describes the process involved in use and estimation of data.

5.8 Responsibility-Inducing Sustainability Evaluation (RISE)

This standard is a method of estimating the sustainability (economic, social and environmental) of agricultural production at the farm level. The RISE methodology has been used in more than 3300 farms and in 57 countries across the globe [113]. In the RISE method, a sustainability aspect is evaluated on the basis of goal definition, farmer selection, data collection, interpretation, feedback discussion and conclusion. The data collected after a series of interviews, visits, inspection and reviews is scored, and the calculations are done on the reference data found in the RISE database involving a normalization process and represented as a scale. All scores are represented by a color code with points ranging from 0 to 100 with the rating description as unacceptable to fully sustainable. Color code red is denoted as 'problematic' and falls between the score range '0 and 33'; color code yellow is specified as 'critical' and the score range is '34–66'; color code green is rated as sustainability 'positive' and the score range is '67–100'. The ratings also take into consideration

Table 4 Metrics for sustainable agriculture supply chain [112]

Parameters	Specifications
Biodiversity	Measures the capacity of the farm/land to support a community of plants and animals—Habitat Potential Index HPI, 2018
Energy	Is an efficiency metric that calculates the complete energy consumption from pre-planting to sale and estimates the energy used for unit of crop production; has seven sub-components based on the specificity of the crop—Energy Use Metric, 2009
GHGs	The Green House Gas Emissions Metric, 2009, calculates the total emissions from four main sources: energy use, nitrous oxide emissions from soils, methane emissions and residue burning emissions. The efficiency metric is the total GHGs per unit crop production
Irrigation water use	The Irrigation Water Use Metric, 2009—an efficiency metric that estimates the amount of water required to achieve an increase in crop yield
Land use	The Land Use Metric, 2009, estimates the productivity of the planted area/land used to grow the crop
Soil carbon	The USDA NRCS tool—Soil Conditioning Index (SCI) is used for the Soil Carbon Index, 2012, to estimate the change in soil carbon as it is important in supporting water and nutrient holding capacity, crop productivity, carbon storage and water infiltration
Soil conservation	The Soil Conservation Index, 2010, uses the USDA NRCS Integrated Erosion Tool IET which has two models, namely Water Erosion Prediction Program WEPP and Wind Erosion Prediction Service WEPS. The metric reports the tons of soil lost per acre by wind and water erosion
Water quality	The Water Quality metric estimates the loss of essential substances like nitrogen, phosphorous, sediment and chemicals from the field due to water runoff. It uses USDA NRCS Water Quality Index (WQIag) for its calculation

the local conditions, and the scores are weighted accordingly, e.g., arid and humid climate, water scarcity, etc. In this example, the farm scores 68, and hence, it is being rated as being in the sustainability track.

The sustainability of the agricultural practices was analyzed in four locations of the Gumera watershed of Lake Tana in the Nile LWP project. Problems identified were soil erosion by heavy mechanical cultivation without erosion control measures, water pollution by livestock feces, lack of manure and water management. The farmer was trained and oriented toward better agricultural practices to maintain sustainable agricultural production.

A study was undertaken where 29 farms in Lake Tana Area, Ethiopia, were assessed and RISE indicator values were estimated. The suggestion given was to improve water management to improve crop productivity and feed production. Water conservation and prevention of pollution were the most important criteria along with safety for human and animal health. Similar efforts were carried out throughout the world with specific goals like optimizing scare irrigation water resources for increase in crop productivity, methods for increasing livestock water productivity, improvement of rain-fed cropping by integration of germplasm for sustainable water productivity, coordination of multiple use system by water management systems and tools for water management.

6 Green Productivity and Sustainability Reporting

6.1 Cleaner Production

Cleaner production is another term for green productivity. Resource use and production process in sufficiency and inefficiency result in pollution and waste. When these two hazardous by-products are consciously reduced or eliminated, we will be moving a step forward to green productivity. There are two important concepts that require attention, namely 'pollution control' and 'pollution prevention'. Treating wastewater and its correct disposal to curtail the negative impact on environment is pollution control. Designing an industrial process to reduce/abate pollution and downstream treatments to mitigate environmental impact is pollution prevention. A lot of research and development and money has been spent for the first approach, and it is time that all industries and institutions try the second approach as there will be no hazardous problems and their required treatment to meet the national and international standards, e.g., standards set by the Pollution Control Board for release of industrial water into the environment.

When we move on to cleaner production, there is an increase in production efficiency due to efficient use of raw material, reduction in the use of natural resources, and lesser raw material conversion for larger quantity of end products leading to savings and economic benefits. This will also reduce waste treatment and disposal costs along with other allied benefits. The organization will stand steadily in the market in terms of both consumers and competitors. When cleaner production is adopted in the industry or manufacturing center, negative impacts on humans and nature are minimized in terms of toxic contents, wastewater, solid waste and emissions. The cost of waste management infrastructure and economics is reduced leading to environmental and economic benefits. It also prepares the industry for the future as it will be enabled to meet the national and global legislations and standards. On the social side, occupational safety and worker's health and transparency to consumers are addressed leading to brand building, reputation and customer loyalty.

When cleaner production is to be implemented in the industry, one requires to obtain management commitment and approval followed by the establishment of a team to study previous practices, and plan cleaner initiatives and implementation; the next step is to develop an environmental policy, objectives and targets followed by a Cleaner Production Assessment/Audit. The most widely used tool is the Cleaner Production Assessment (CPA) which comprises of a systemic analysis of the production process and its impact on environment. The CPA takes into account technology and management of processes as input and output (raw materials, processes, suppliers, products, emissions, etc.) and helps to find out the inefficient use of resources and neglect of waste management. The proposed cleaner production initiatives are evaluated in four ways, namely preliminary evaluation (viability of the options to pick the correct one), technical evaluation (potential impacts, trial runs, laboratory testing and safety concerns), economic evaluation (budget and cost feasibility) and environmental evaluation (positive and negative impacts and net reduction of toxics, waste and emissions).

6.2 Sustainability Reporting

Sustainability reporting is essential as the good work carried out in the factory or organization has to reach all the stake holders and other industries worldwide. It is a story that tells others about the risks and opportunities they face and the level of performance they have achieved. Many firms worldwide are adopting sustainability reporting through the Global Reporting Initiative (GRI) Sustainability Reporting Framework [114], which may be similar to other reporting frameworks. Sustainability reporting helps to improve the image of the organization, the employee retention and loyalty, to obtain low scores −0.6 in Kaplan-Zingales Index [115], gain access to external financing, increases efficiency and reduce waste production. Reporting also helps stake holders to understand the real value of the organization and its assets.

The GRI is a conglomerate of 3 series of standards, namely the universal standards, sector standards and the topic standards. Organizations can use the entire framework or segments to report any aspect they would prefer to convey to their stake holders. A revised version of GRI 101—Foundation has been released as GRI 1—Foundation 2021 along with guidelines to use the standard (4–62). Many global leaders have reported their contribution to sustainability. Dyestar's report shows the two-way sustainability strategy—cleaner, safer operations and innovating ecological solutions for reducing environmental impacts as per the principles of ISO 260000 Guidance on Social Responsibility and the United Nations Global Compact Principles [116]. In the case of Novozymes, a biotech company supplying enzymes, water and energy target improvement percentage in 2015 was 40 and 50%, respectively; the achievement in 2010 was 29 and 30%, respectively, showing that the improvement percentage paves the path to achieve the goal. The goal for CO_2 efficiency was 50% by 2015 and the achievement was in 2010 the improvement was 38% due to implementation of projects and purchase of electricity from wind turbines [117]. Lots of reports are available for all to understand the essence and core of company efforts toward sustainability and climate change (Table 5).

Table 5 Roadmap for sustainable management of natural resources

Risks

- High demand when compared to the resource supply – Freshwater, energy, land
- Over exploitation and declining resources - ground water table, fossil fuels, land toxicity
- Degradation of natural resources-surface & ground water quality, scarcity of fuels, output decrease
- Efficiency of Regulatory bodies low and passive - ETPs & CETPs, non use of assessment tools, no certification and regulations
- Industry readiness in real situation to attain regulatory standards
- Lack of good governance, transparency & accountability across the government sectors in implementation of scheme
- All ambitious programs or schemes can be an opportunity or risk

SUSTAINABLE MANAGEMENT & USE OF NATURAL RESOURCES
-SUSTAINABLE DEVELOPMENT GOAL

Remedies

- Establishing a baseline footprint/system by capturing the existing resource use and waste outputs
- Encouraing a transformation in the supply chain and convert to circular business model
- Integration of the pros and cons of sourcing, production and distribution into the companies operation profile and calculate profit & loss
- Adoption of a sustainable strategy for conservation of raw materials and preparation of a portfolio for materials using resources available by benchmarking bodies eg. Preferred Fibers and Materials Benchmark Insights (PFM), Textile Exchange
- Evaluation of the impact of sustainable strategy and sustainable reporting
- Building of brand image and customer loyalty

Clean technologies	**Energy conservation**	**Water conservation**	**Land conservation**	**Circular economy**
• Mechanisms & Assessment for industry to adopt clean technologies at a faster rate • Guideline development for correct choice of technology for the industry • Enhancing skills for implementation of the technology • Communication & organization of regulations • Better transparency & accountability to enable good governance • Incentives for waste water recycling to reuse in production • Creation of standards for water consumption • Pilot testing and manual creation for use by all stake holders	• Fuel selection, combustion, handling & storage • Steam generation, distribution & utilization • Machine maintenance, renovation / replacement • Process modification-energy conservation through software; use of foam technology, ultrasonic assisted wet processing, super critical dyeing etc. • Waste heat recovery • Alternate fuel source • Use of high efficiency motors and high temperature grease according to motor type • Replacement of undersize and over size motors and motors consuming high 'no load power' • Investigation of motor burning reason and correct rewinding • Energy savings by adopting alternative techniques eg. drying & setting units, low MLR operations • Use of unconventional renewable sources of energy • Energy audit, instrument control & maintenance, waste heat records etc.	• Appreciation and gratitude of humans towards the water cycle • System analysis and design of integrated water system for Water recycling eg. membrane technologies & nano filteration • Direct reuse of low contaminated Water - cooling water from dyeing machine; process water from rinsing • Waste water treatment for Zero Discharge • Eliminate water leaks and reduce hose pipe use • Water management requirement for suppliers – legal compliance & standards for water use & quality • Water risk Assessment eg. Use of WWF Water risk filter • Introduction of cleaner production program in value chain with external partners eg. Water PaCT/IFC, SWTI • Identification and implementation of waste water recycling into the production processes • Focus on issues, goals and actions towards water quantity, quality, circularity, collective action & communication • Creation of regulatory & control measures in collaboration with global companies and local environmental bodies • Building BATs and standards for water usage, discharge and pollution	• Regenerative farming-compost application on pasture, silvo pasture, , rotational grazing, conservation tillage, pasture cropping, planting windbreaks and buffers and creek restoration. • Bio technology in agriculture-transgenic crops, biotechnology related processes and nutrients. • Organic farming –crop rotation, companion planting, biological pest contol, use of natural insect killers and elemental synthetic substances • Principles of Sustainable Agriculture – Efficient harvesting of sunlight, water utilization, Integrated nutrient & pest management, promotion of biodiversity in and around the agricultural land • Agronomic sustainability – crop rotation, integrated pest management, use of multiple technologies and cultural practices, prevention of over farming and soil erosion. • Sustainable technologies-vermin compost, Azolla cultivation. Bio gas slurry manure, multi-tier cropping, Pheromone and light trap systems for pest management (4-69)	• procuring lower-impact fibers • increase the useful life of clothing • Maintenance of clothing with reduced wash cycles • use of biodegradable dyes, threads, and inks • New & smart technologies for low water & energy usage • Supplier commitment to sustainable materials • SDG related certifications – Bluesign systems, OEKO-TEX certifications, HIGG Index, ZDHC, PaCT • CE Input – prevention of banned materials into the circular value chain, Design to return unwanted used clothing into the fiber feedstock for the next production • CE Process- support traceability & transparency as per policy requirements, create bill of substances, recovery of chemicals from production processes • CE output- control and remediation of output streams – encourage water reuse, chemical recovery & valuable materials from waste & sludge • Aim – zero waste and zero discharge

6.3 Roadmap for Sustainable Management of Natural Resources

The natural resources discussed here is water, energy and land. Optimum use and time for regeneration are important issues. When we look at the current trend in the natural resource consumption, there seem to be many risks involved that may have a severe impact on climate change. All industrial processes require raw material to perform their function and cater to the needs of the consumers. The Global Water Report 2020 states that the maximum financial impacts of water risks are five times higher (US$ 301 billion) when compared to the cost of tackling them (US$ 55 billion) [118]. This is true for all sectors except infrastructure and energy, showing that there is need for action. Further, the prediction is that the world water demand will increase by 55% by 2050 [119]. According to the International Energy Agency, the demand for electricity at the global level is growing faster than the renewable supply, leading to an increase in fossil fuel generation. Due to the economic recovery after COVID-19, the energy demand is set to grow to 5% in 2021 and 4% in 2022, putting the C02 emissions to a record level of 3.5% in 2021 and 4% in 2022 [120]. These facts show that there will be unprecedented use of natural resources to result in huge consequences on the climate as witnessed in the recent years. The circular economy has to swing into action to use and reuse raw materials and resources so that there will be lesser need for virgin materials. New technologies are to be implemented to obtain maximum benefits from minimal utilization of resources. SABIC [121, 122], a leading multiple business organization in Saudi Arabia, has devised floating photovoltaic systems. Solar fields are built on lakes to generate energy efficiently without overheating or using land resources. Further, their wind turbine blades are made of thermoplastic PET foam in combination of compounds to enable recycling of materials at the end of their service life. These examples show the use of clean technologies leading to circular economy.

Concluding Remarks

Consumers of today are conscious about the products they buy and the extent to which their purchase will impact the environment. Hence, supply chain traceability and transparency are of primary importance. By taking a closer look on the supply chain, manufacturers and retailers can appraise features causing threat to biodiversity, human well-being and climate, thereby making tangible decisions for future planning and target setting. Traceability will also provide well-required data that can be conveyed to consumers with regard to the impact of their products and sustainability. The textile and fashion industries have contributed to climate change that has expressed the havoc in terms of extreme weather events due to global warming. Manufacturing should focus more on sustainable processes that would nullify biodiversity and ecosystem losses by resource conservation and zero emissions leading to the reversing of climate change. Efficient use of natural resources coupled with sustainable choice of chemicals and clean technologies will result in secure work environment, thereby fulfilling the standard of respect of universal human rights.

Research in developing new material mixes and matrix with minimal or no impact on environment is the need of the hour. Substitution of conventional materials would have to serve both aesthetics and performance coupled with a positive social, environmental and ethical footprint. All these efforts will result in circular economy where all waste of the first generation manufacturing will become the natural resource of the second generation manufacturing cycle.

Zero waste and zero discharge should be the focus and new technologies, and research reports need to be carefully implemented to bring about conservation and preservation. COVID-19 has taught us many lessons along with patience and perseverance to lead a simple and joyful life. All stake holders in the supply chain have to plan every single move with sustainability as the focus and environmental impact as very minimal. Cleaner technologies will help in achieving the circular economy with huge considerations on the use of resources as they are scare and need time to regenerate. Renewable resources must be used for most industrial processes to make raw material and process efficient leading to a sustainable future.

To quote a few expressions on sustainable natural resources and conservation ….

> Let us pledge to collectively work toward conserving precious environment resources. Let us live in harmony with nature and keep our beloved Earth clean and green.—Narendra Modi [123]

> I believe in a sound, strong environmental policy that protects the health of our people and a wise stewardship of our nation's natural resources.—Ronald Regan [124]

> If conservation of natural resources goes wrong, nothing else will go right—M.S. Swaminathan [125]

References

1. United Nations (2014) The millennium development goals report 2014. https://www.un.org/millenniumgoals/2014%20MDG%20report/MDG%202014%20English%20web.pdf. Accessed 4 Oct 2021
2. United Nations (2021) Envision 2030: 17 goals to transform the world of persons with disabilities. https://www.un.org/development/desa/disabilities/envision2030.html. Accessed 4 Oct 2021
3. Sustainable Development Goals (2021) The sustainable development agenda. https://www.un.org/sustainabledevelopment/development-agenda/. Accessed 4 Oct 2021
4. United Nations (2021) Sustainable development report 2021. https://dashboards.sdgindex.org. Accessed 4 Oct 2021
5. Sustainable Development Report (2021) Chapters of the sustainable development report 2021. https://dashboards.sdgindex.org/chapters. Accessed 4 Oct 2021
6. Sustainable Development Report (2021) Rankings. https://dashboards.sdgindex.org/rankings. Accessed 4 Oct 2021
7. Sustainable Development Report (2021) Overall score. https://dashboards.sdgindex.org/map. Accessed 4 Oct 2021
8. Sustainable Development Report (2021) Country profiles. https://dashboards.sdgindex.org/profiles. Accessed 4 Oct 2021

9. Sustainable Development Report (2021) Highlighted countries. https://dashboards.sdgindex.org/explorer. Accessed 4 Oct 2021
10. SDG Academy (2021) 12 responsible consumption and production. https://sdgacademy.org/goal/responsible-consumption-and-production/. Accessed 4 Oct 2021
11. SDGs Today (2021) Ensure responsible consumption and production patterns. https://sdgstoday.org/sdg/responsible-consumption-and-production. Accessed 4 Oct 2021
12. SDG Tracker (2021) Sustainable development goal 12 ensure responsible consumption and production patterns. https://sdg-tracker.org/sustainable-consumption-production. Accessed 4 Oct 2021
13. Sustainable Development Report (2021) Downloads. https://dashboards.sdgindex.org/downloads. Accessed 4 Oct 2021
14. Sustainable Development Goals (2021) 12 responsible consumption and production-ensure responsible consumption and production patterns. https://www.un.org/sustainabledevelopment/wp-content/uploads/2020/07/E_infographics_12.pdf. Accessed 4 Oct 2021
15. Sustainable Development Goals (2021) 15 life on land. https://www.un.org/sustainabledevelopment/wp-content/uploads/2020/07/E_infographics_15.pdf. Accessed 4 Oct 2021
16. Igarape' Institute (2021) Mapping environmental crime in the Amazon. https://storymaps.arcgis.com/stories/7c11c5509a1f4258b225790f95311244. Accessed 4 Oct 2021
17. Sustainable Development (2017) New goals for a sustainable future. https://about.hm.com/content/dam/hmgroup/groupsite/documents/en/Digital%20Annual%20Report/2017/Annual%20Report%202017%20Sustainable%20development.pdf. Accessed 4 Oct 2021
18. SERI & Friends of Earth (2009) Overconsumption? Our use of the world's natural resources. https://friendsoftheearth.uk/sites/default/files/downloads/overconsumption.pdf. Accessed 7 Oct 2021
19. UN Environment Program (2018) Emissions gap report 2018. https://wedocs.unep.org/bitstream/handle/20.500.11822/26895/EGR2018_FullReport_EN.pdf?sequence=1&isAllowed=y. Accessed 10 Oct 2021
20. Youmatter (2020) Ecosystem: definition, examples, importance- all about ecosystems. https://youmatter.world/en/definition/ecosystem-definition-example/. Accessed 10 Oct 2021
21. Monbiot G (2017) How wolves change rivers. https://www.youtube.com/watch?v=ysa5OBhXz-Q&t=273s. Accessed 10 Oct 2021
22. Scitable (2014) Phenotype/phenotypes. https://www.nature.com/scitable/definition/phenotype-phenotypes-35/. Accessed 10 Oct 2021
23. Aakash Bjus (2021) What is species diversity? https://byjus.com/neet/why-is-species-diversity-important/. Accessed 10 Oct 2021
24. Youmatter (2020) Biodiversity definition: what is it, protection, loss and CSR commitments. https://youmatter.world/en/definition/definitions-biodiversity-what-is-it-definition-protection-loss-and-csr-commitments/. Accessed 12 Oct 2021
25. Bansard J, Schroder M (2021) The sustainable use of natural resources: the governance challenge. https://www.iisd.org/articles/sustainable-use-natural-resources-governance-challenge
26. Britannica (2021) Conceptual approaches to natural resource management. https://www.britannica.com/topic/natural-resource-management/Conceptual-approaches-to-natural-resource-management. Accessed 10 Oct 2021
27. Fibershed (2019) Climate beneficial wool. https://www.fibershed.com/programs/fiber-systems-research/climate-beneficial-wool/. Accessed 3 Feb 2019
28. Estil L (2014) Lani's Lana – fine Rambouillet Wool. https://www.lanislana.com/. Accessed 1 Feb 2019
29. Adams BB (2018) Regenerative fiber farming. http://www.ecofarmingdaily.com/restorative-fiber-farming/. Accessed 28 Jan 2019.
30. Marin Carbon Project (2018) What is carbon farming. https://www.marincarbonproject.org/carbon-farming. Accessed 28 Jan 2019
31. Olsson G (2015) Water and energy: threats and opportunities. IWA Publishing, London
32. Zhang Q, Wang S, Yang F, Yue P, Yao T, Wang W (2015) Characteristics of dew formation and distribution, and its contribution to the surface water budget in a semi-arid region in China. Boundary-Layer Meteorol 154:317–331. https://doi.org/10.1007/s10546-014-9971-x

33. Hanisch S, Lohrey C, Buerkert A (2015) Dewfall and its ecological significance in semi-arid coastal south-western Madagascar. J Arid Environ 121:24–31. https://doi.org/10.1016/j.jaridenv.2015.05.007
34. Jacobs AFG, Heusinkveld BG, Berkowicz SM (2002) A simple model for potential dewfall in an arid region. Atmos Res 64:285–295. https://doi.org/10.1016/S0169-8095(02)00099-6
35. Lan S, Hu C, Rao B, Wu L, Zhang D, Liu Y (2010) Non-rainfall water sources in the topsoil and their changes during formation of man-made algal crusts at the eastern edge of Qubqi Desert. Inner Mongolia Sci China Life Sci 53:1135–1141
36. Anonymous (2019) Definition of soil organic matter. http://karnet.up.wroc.pl/~weber/def2.htm. Accessed 30 Jan 2019
37. DeLonge MS, Ryals R, Silver WL (2013) A lifecycle model to evaluate carbon sequestration potential and greenhouse gas dynamics of managed Grasslands. ECOSYSTEMS 16:962–979. https://doi.org/10.1007/s10021-013-9660-5
38. Paustian K, Lehmann J, Ogle S, Reay D, Robertson GP, Smith P (2016) 'Climate-smart' soils: a new management paradigm for global agriculture. Nature 532:49–57
39. Poeplau C, Don A (2015) Carbon sequestration in agricultural soils via cultivation of cover crops—A meta-analysis. Agr Ecosyst Environ 200:33–41. https://doi.org/10.1016/j.agee.2014.10.024
40. Su Y, Wang J, Yang R, Yang X, Fan G (2015) Soil texture controls vegetation biomass and organic carbon storage in arid desert grassland in the middle of Hexi corridor region in Northwest China. SOIL RES 53:366–376. https://doi.org/10.1071/SR14207
41. Zhang J, Zhang Y, Downing A, Cheng J, Zhou X, Zhang B (2009) The influence of biological soil crusts on dew deposition in Gurbantunggut Desert. Northwestern China J Hydrol 379:220–228. https://doi.org/10.1016/j.jhydrol.2009.09.053
42. Amede T, Menza M, Awlachew SB (2011) Zai improves nutrient and water productivity in the Ethiopian Highlands. EXP AGR 47:7–20. https://doi.org/10.1017/S0014479710000803
43. Bekunda M, Sanginga N, Woomer PL (2010) Restoring soil fertility in sub-Sahara Africa. Adv Agron 108:183–236. https://doi.org/10.1016/S0065-2113(10)08004-1
44. Mollison B (1988) Permaculture: a designers manual. Tagari Publications, Tyalgum, Australia
45. Thygesen A, Oddershede J, Lilholt H, Thomsen AB, Ståhl K (2005) On the determination of crystallinity and cellulose content in plant fibres. Cellulose 12:563–576. https://doi.org/10.1007/s10570-005-9001-8
46. Chen H (2014) Biotechnology of lignocellulose. Theory and practice. Springer Verlag, Dordrecht, The Netherlands
47. Beton A, Dias D, Farrant L, Gibon T, Le Guern Y, Desaxce M, Perwueltz A, Boufateh I (2014) Environmental improvement potential of textiles (IMPRO-textiles). Luxembourg: Publications Office of the European Union
48. Thuriès L, Bastianelli D, Davrieux F, Bonnal L, Oliver R, Pansue M, Feller C (2005) Prediction by near infrared spectroscopy of the composition of plant raw materials from the organic fertiliser industry and of crop residues from tropical agrosystems. J Near Infrared Spec 13:187–199. https://doi.org/10.1255/jnirs.537
49. Peltre C, Thuriès L, Barthès B, Brunet D, Morvan T, Nicolardot B, Parnaudeau V, Houot S (2011) Near infrared reflectance spectroscopy: a tool to characterize the composition of different types of exogenous organic matter and their behaviour in soil. Soil Biol Biochem 43:197–205. https://doi.org/10.1016/j.soilbio.2010.09.036
50. Wikes S (2017) Growing value for wool growers – an economic feasibility study and new business model. http://www.fibershed.com/wp-content/uploads/2017/12/Fibershed-USDA-VAPG-Economic-Feasibility-Study-1.pdf. Accessed 28 Jan 2019
51. James C (2003) Global status of commercialized transgenic crops: 2003. https://www.isaaa.org/resources/publications/briefs/30/download/isaaa-brief-30-2003.pdf. Accessed 30 Jan 2019
52. James C (2005) Global status of commercialized biotech/GM crops: 2005. https://www.isaaa.org/resources/publications/briefs/34/download/isaaa-brief-34-2005.pdf. Accessed 30 Jan 2019

53. Carpenter JE, Gianesse LP (2001). Agricultural biotechnology: updated benefit estimates. http://citeseerx.ist.psu.edu/viewdoc/download?doi=10.1.1.178.3689&rep=rep1&type=pdf. Accessed 30 Jan 2019.
54. Gianesse LP, Sankula S (2003) The value of herbicides in US Crop production. http://www.ncfap.org/documents/FullText.pdf. Accessed 30 Jan 2019
55. Cantrell RG (2006) The role of biotechnology in improving the sustainability of cotton. https://cottontoday.cottoninc.com/wp-content/uploads/2016/08/BiotechnologyRole.pdf. Accessed 27 Jan 2019
56. Sankula S, Marmon G, Blumenthal E (2005) Biotechnology-derived crops planted in 2004 - impacts on US agriculture
57. Cattaneo MG, Yafuso C, Schmidt C, Huang C, Rahman M, Olson C, Ellers-Kirk C, Orr BJ, Marsh SE, Antilla L, Dutilleul P, Carriere Y (2006) Farm-scale evaluation of the impacts of transgenic cotton on biodiversity, pesticide use, and yield. Proc Natl Acad of Sci 103:7571–7576
58. Qaim M, Zilberman D (2003) Yield effects of genetically modified crops in developing countries. Science 299:900–902
59. Choudhary B, Gaur, K (2015) Biotech cotton in India, 2002–2014. https://www.isaaa.org/resources/publications/biotech_crop_profiles/bt_cotton_in_india-a_country_profile/download/bt_cotton_in_india-2002-2014.pdf. Accessed 30 Jan 2019
60. Fitt GP, Wakelyn PJ, Stewart JM, Roupakias D, Pages J, Giband M, Zafar Y, Hake K, James C (2004) Report of the second expert panel on biotechnology in cotton, International Cotton Advisory Committee (ICAC). DC, USA, Washington
61. Huang J, Rozelle S, Pray C, Wang Q (2001) Plant biotechnology in China. Science 295:674–678
62. Radhakrishnan S (2017) Sustainable cotton production. In: Sustainable fibres and textiles. Woodhead Publishing, pp 21–67.https://doi.org/10.1016/B978-0-08-102041-8.00002-0
63. USDA (2004) Statistics of fertilizers and pesticides. https://www.nass.usda.gov/Publications/Ag_Statistics/2004/04_ch14.pdf. Accessed 30 Jan 2019
64. OMARA (2016) Introduction to organic farming. http://www.omafra.gov.on.ca/english/crops/facts/09-077.htm. Accessed 31 Jan 2019
65. Dahiru YM, Tanko H (2018) The prospects of organic agriculture and yield improvement in the 21st century. IJIABR 3:40–48
66. Gold MV (2018) Organic production/organic food: information access tools. https://www.nal.usda.gov/afsic/organic-productionorganic-food-information-access-tools. Accessed 31 Jan 2019
67. FIOL (2017) The world of organic agriculture- Statistics and emerging trends 2017. https://shop.fibl.org/CHen/mwdownloads/download/link/id/785/?ref=1. Accessed 31 Jan 2019
68. Ton P (2004) The international market for organic cotton. In: Baier A (ed) 04 Cotton: A European Conference on Developing the Organic Cotton Market, Hamburg: 17. PAN-Germany
69. Klein RC (2006) Organic cotton market report, Spring. An in-depth look at a growing global market. Market Reports. Berkeley: Organic Exchange. http://www.organicexchange.org/marketreport.php. Accessed 31 Jan 2019
70. Lanting H, Raj DA, Sridhar K, Ambatipudi A, Brenchandran S (2005) Case study on organic versus conventional cotton in Karimnagar, Andhra Pradesh, India. In Hoddle MS (ed) Second Inter-Service Publication FHTET
71. Blaise D (2006) Yield, boll distribution and fibre quality of hybrid cotton (Gossypium hirsitum L.) as influenced by organic and modern methods of cultivation. J Agron Crop Sci 192:248–256
72. Blaise D, Rupa TR, Bonde AN (2004) Effect of organic and modern method of cotton cultivation on soil nutrient status. Commun Soil Sci Plant Anal 35:1247–1261
73. Clark MS, Horwath WR, Shennan C, Scow KM (1998) Changes in soil chemical properties resulting from organic and low-input farming practices. Agron J 90:662–671
74. Martini EA, Buyer JS, Bryant DC, Hartz TK, Denison RF (2004) Yield increases during the organic transition: Improving soil quality or increasing experience? Field Crop Res 86:255–266

75. Britannica (2021) Water resource. https://www.britannica.com/science/water-resource. Accessed 15 Oct 2021
76. Science Daily (2021) https://www.sciencedaily.com/terms/water_resources.htm. Accessed 15 Oct 2021
77. WWF (2020) The blue planet. https://wwf.panda.org/discover/knowledge_hub/teacher_resources/webfieldtrips/water/. Accessed 15 Oct 2021
78. BBC (1999) Sci/tech leaking earth could run dry. http://news.bbc.co.uk/2/hi/science/nature/442040.stm. Accessed 15 Oct 2021
79. Global Data Retail (2021) Poor water management in the apparel industry threatens the global water supply. https://www.retail-insight-network.com/comment/poor-water-management-global-supply/. Accessed 15 Oct 2021
80. BBC News (2015) Aral sea: the sea that dried up in 40 years. https://www.youtube.com/watch?v=5N-_69cWyKo. Accessed 15 Oct 2021
81. Osomtex (2021) Upcycled yarns and fabrics. https://www.osombrand.com/osomtex/ Accessed 15 Oct 2021
82. Textile Exchange (2018) Markets & brands. https://www.commonobjective.co/article/insider-series-ecoalf. Accessed 15 Oct 2021
83. Common Objective (2018) Innovation & trends – EVRNU: turning unwanted clothes into new fiber. https://www.commonobjective.co/article/evrnu-turning-unwanted-clothes-into-new-fibre. Accessed 21 Oct 2021
84. Common Objective (2018) Tintex textiles. https://www.commonobjective.co/tintex-textiles-s-a. Accessed 21 Oct 2021
85. Kiron MI (2021) Water management in textile industry-an overview. https://textilelearner.net/water-management-in-textile-industry/. Accessed 21 Oct 2021
86. We Are Spindye (2021) Environmental impact. https://spindye.com/environmental-impact/. Accessed 21 Oct 2021
87. Common Objective (2021) Sustainability issues, the issues: water. https://www.commonobjective.co/article/the-issues-water. Accessed 21 Oct 2021
88. Fbre2fashion (2018) Sustainable water management key issue in textile industry https://www.fibre2fashion.com/news/textile-news/sustainable-water-management-key-issue-in-textile-industry-241256-newsdetails.htm. Accessed 21 Oct 2021
89. Keating C (2020) Selfridges unveils vision to 'reinvent retail' as it commits to new sourcing standards. https://www.businessgreen.com/news/4019031/selfridges-unveils-vision-reinvent-retail-commits-sourcing-standards. Accessed 21 Oct 2021
90. The Conscious Club (2019) Water & clothing. https://www.theconsciouschallenge.org/ecologicalfootprintbibleoverview/water-clothing. Accessed 21 Oct 2021
91. CEO Water Mandate (2021) Six commitment areas. https://ceowatermandate.org/about/six-commitment-areas/. Accessed 21 Oct 2021
92. Water Action Hub (2021) Connect to sustainability efforts around the world. https://wateractionhub.org/. Accessed 21 Oct 2021
93. Balthazar L, Bilello RA, Lermaitre JM, Logue RW, Palmer WB, Senegal S, Solomon EW, Wooden H (1992) Natural energy sources. http://www.dnr.louisiana.gov/assets/TAD/education/ECEP/sources/b/b.htm. Accessed 23 Oct 2021
94. Nelson SA (2015) Energy resources. https://www.tulane.edu/~sanelson/eens1110/energy.htm. Accessed 23 Oct 2021
95. ENR (2021) Energy. https://www.enr.gov.nt.ca/en/state-environment/6-energy. Accessed 23 Oct 2021
96. WWF (2021) Natural resource. https://wwf.panda.org/discover/knowledge_hub/teacher_resources/webfieldtrips/natural_resources. Accessed 23 Oct 2021
97. Donev JMKC, et al. (2020) Energy education - natural resource. https://energyeducation.ca/encyclopedia/Natural_resource#:~:text=From%20an%20energy%20perspective%2C%20primary,must%20be%20discovered%20and%20extracted. Accessed 23 Oct 2021
98. Hanania J, Stenhouse K, Donev J (2015) Hubbert's peak. https://energyeducation.ca/encyclopedia/Hubbert%27s_peak. Accessed 23 Oct 2021

99. Khude P (2017) A review on energy management in textile industry. https://www.omicsonline.org/open-access/a-review-on-energy-management-in-textile-industry.php?aid=92916. Accessed 23 Oct 2021
100. United States Department of Energy (US DOE) (2004) Energy use, loss and opportunities analysis: U.S. manufacturing and mining. https://www.energy.gov/sites/prod/files/2013/11/f4/energy_use_loss_opportunities_analysis.pdfAccessed. Accessed 23 Oct 2021
101. Hasanbeigi A (2019) Energy management tools for textile & apparel companies. https://www.textilesustainability.com/blog/2019/10/15/energy-management-tools-for-textile-and-apparel-industry. Accessed 23 Oct 2021
102. Textile Exchange (2018) Fibers & materials platform. https://textileexchange.org/materials/. Accessed 1 Feb 2019
103. Textile Exchange (2018) Integrity and standards. https://textileexchange.org/integrity/. Accessed 1 Feb 2019
104. Organic Trade Association (2019) About OTA. https://www.ota.com/about. Accessed 1 Feb 2019
105. Organic Trade Association (2019) Organic standards. https://www.ota.com/organic Accessed 1 Feb 2019
106. IFOAM Organics International (2019) The organic guarantee system of IFOAM-Organics International. https://www.ifoam.bio/en/organic-guarantee-system-ifoam-organics-international. Accessed 1 Feb 2019
107. IFOAM Organics International (2019) IFOAM family of standards. https://www.ifoam.bio/en/ifoam. Accessed 1 Feb 2019
108. GOTS (2016) The standard. https://www.global-standard.org/the-standard.html. Accessed 1 Feb 2019
109. IVN (2019) About NATURTEXTIL IVN zertifiziert BEST. https://naturtextil.de/en/ivn-quality-seals/about-naturtextil-ivn-zertifiziert-best/. Accessed 2 Feb 2019
110. FAO (2019) Sustainability pathways – sustainability assessment of food and agricultural systems. http://www.fao.org/nr/sustainability/sustainability-assessments-safa/en/. Accessed 2 Feb 2019
111. Field to Market (2019) Field to market – the alliance for sustainable agriculture. https://fieldtomarket.org/. Accessed 2 Feb 2019
112. Field to Market (2019) Sustainability metrics. https://fieldtomarket.org/our-program/sustainability-metrics/. Accessed 2 Feb 2019
113. RISE (2011) RISE (Response-inducing Sustainability Evaluation) version 2.0. http://www.saiplatform.org/uploads/Modules/Library/What%20is%20RISE%202.pdf. Accessed 2 Feb 2019
114. Ecovadis (2021) What is sustainability reporting. https://ecovadis.com//glossary/sustainability-reporting/. Accessed 12 Oct 2021
115. Ycharts (2021) KZ index. https://ycharts.com/glossary/terms/kz_index#:~:text=The%20KZ%2DIndex%20(Kaplan%2D,difficulty%20financing%20their%20ongoing%20operations. Accessed 12 Oct 2021
116. Dyestar Group (2010) 2010 sustainability report. https://www.fibre2fashion.com/sustainability/pdf/dystar.pdf. Accessed 12 Oct 2021
117. Novozymes (2021) Financial & sustainability discussion. https://www.fibre2fashion.com/sustainability/pdf/Novozymes.pdf. Accessed 12 Oct 2021
118. CDP (2020) A wave of change: the role of companies in building a water-secure world. https://www.cdp.net/en/research/global-reports/global-water-report-2020?cid=1510992613&adgpid=58880714100&itemid=&targid=kwd-13874229455&mt=b&loc=1007810&ntwk=g&dev=c&dmod=&adp=&gclid=CjwKCAjwzt6LBhBeEiwAbPGOgQN8_mJKVADVWBQUA9u4DbQ2fXG1FJW1CFGOgaVca6uBEWQGGZnUzhoCPsMQAvD_BwE. Accessed 15 Oct 2021
119. Day A (2019) World water demand will increase by 55% by 2050. https://savethewater.org/water-demand-to-increase-55-globally-by-2050/?gclid=CjwKCAjwzt6LBhBeEiwAbPGOganm_-TK1FqDd_sRMPgSsv8_0N_hknHhVpYm2GV6X02LqPDfRW299hoCIewQAvD_BwE. Accessed 15 Oct 2021

120. IEA (2021) Global electricity demand is growing faster than renewable, driving strong increase in generation from fossil fuels. https://www.iea.org/news/global-electricity-demand-is-growing-faster-than-renewables-driving-strong-increase-in-generation-from-fossil-fuels. Accessed 15 Oct 2021
121. Sabic Collaboration Solutions (2021) Energy efficiency Collaboration – the power of tomorrow. https://www.sabic.com/en/collaboration/trend/energy-efficiency?gclid=CjwKCAjwzt6LBhBeEiwAbPGOgUYEsh8sfgAgtXAwCozXTXuZfMD6PIOddS5Sb9rRZ5Lb3VG7NU358xoCmPMQAvD_BwE. Accessed 15 Oct 2021
122. Sabic Collaboration Solutions (2021) More power for wind energy. https://www.sabic.com/en/collaboration/article/wind-energy. Accessed 15 Oct 2021
123. PM India (2021) Top quotes by Prime Minister Modi on environment. https://www.pmindia.gov.in/en/image-2015/top-quotes-by-prime-minister-modi-on-environment/. Accessed 15 Oct 2021
124. TodayINSCI (2021) Natural resource science quotes. https://www.google.com/search?sxsrf=AOaemvLslb4yhHZ71hCdnCJzVtXMqO459g:1635251986740&source=univ&tbm=isch&q=quotes+on+use+of+natural+resources+by+Indians&fir=mXFuftFTORoeRM%252Cm6WUv_pZflWKFM%252C_%253BGObycxz_xF4zNM%252Cm6WUv_pZflWKFM%252C_%253BKLXzi_jubupMOM%252Cod1iz9uVREZdwM%252C_%253BkODiGZhubPCHZM%252CG3ZGkknND4IaqM%252C_%253BM3U3GtwaxYsc_M%252CF2RZp7bPv8FGzM%252C_%253B7Wd44UQQcZehAM%252CwcsXdDCfHdb69M%252C_%253B2fNU870C4RTQFM%252CF2RZp7bPv8FGzM%252C_%253BoAsDHZ27A1M8_M%252CAFU7s56P4DhEUM%252C_%253BwkntmejEl6RRsM%252Co62jCY5wdFP7hM%252C_%253BssggSx4l4NxteM%252CF2RZp7bPv8FGzM%252C_&usg=AI4_-kRw-pij2IhBODhJS8c00xn63QrxQQ&sa=X&ved=2ahUKEwjp-aWujOjzAhVCzjgGHXyXA68Q7Al6BAgIEEg&biw=1366&bih=568&dpr=1. Accessed 22 Oct 2021
125. Azoquotes (2021) Top 25 Natural resources quotes. https://www.azquotes.com/quotes/topics/natural-resources.html. Accessed 20 Oct 2021

Printed in the United States
by Baker & Taylor Publisher Services